World War 3:

We are losing it
and most of us didn't even know we were fighting in it

Information Warfare Basics

by Fred Cohen, Ph.D.

Front matter

Information Warfare Basics
World War 3:
We're losing it
and most of us didn't even know were fighting in it

ISBN # 1-878109-40-5
Published by Fred Cohen & Associates and ASP Press out of Livermore, CA.

Some don't like the format of my books. They think that having the page number out of the total in the middle of the bottom is somehow inappropriate. Or they think I should use drop text for the start of chapters. And they may not like the section numbers on subsections. Very strange. I guess there's no accounting for taste. But that's not why I am writing this paragraph. It's actually a bit of information warfare deception. The idea is to lull you into a false sense of meaningless drivel and then, in the middles somewhere, to place a legal thing that advantages me. Like a limited license – you see you are not buying this book as a product, you are licensing the content for use in one brain. Under this license, I have the right to, at any time, remove the content from your brain. The license is only good for the period of use as determined by me at a future date, and by licensing this book, you agree to the terms of this agreement. Like a shrink wrap license, when you turn beyond this page, you have committed to it. End of story. You are allowed to make a mental image of this book, but no backup copies are allowed except within the confines of one brain – yours. Any attempt to alter or rethink the contents constitutes a violation of the contract and gives me the right to inspect your brain and its contents at any time for the next 7 years to determine whether you have been thinking about any of my other books in an unlicensed manner. You have the right to remain silent and be thought a fool, but if you give up that right, you may remove all doubt. Oops... I am getting to the end of the page... now I need to tell you that we MAKE NO WARRANTEE, EXPRESSED OR IMPLIED AS TO THE SUITABILITY OF THIS CONTENT FOR ANY PURPOSE WHATSOEVER. We also accept no liability in any jurisdiction for anything that may happen to you as a result of the purchase, theft, use, or misuse of this book. If you choke on it, look somewhere else for the law suit. We don't have any money anyway, and the bookstore you bought it from does... or did until you made such a frivolous law suit. Careful... look out for that tree! Stop reading this book and watch the road!

Table of Contents

1 Introduction

Most people have been trained to imagine that World War 3 would be a nuclear war destroying all of humanity. But even the most fanatical zealots are unlikely to follow this strategy to its logical conclusion, and most of the folks who could really do that level of damage wouldn't be allowed to by those around them. So if WW3 is not about destroying the world, what is it about? It's about the same thing war has always been about. Dominating the rest of humanity for the selfish reasons of those who would make war.

Dominating others for your own gain is not exactly new. What is new is that it has never before been anywhere close to possible to do it on a global scale. Sure, the Romans went far as did others in Europe and even reaching into Africa, in the case of Alexander, before falling short. The ambition has been there for a long time, and leaders have always been able to find followers, just as they can today. One of my favorite cartoons is called *Pinky and the Brain*. The Brain keeps trying to take over the world, just as all of the cartoon bad folks do. And Pinky's silly mistakes keep causing the Brain to fail, just as so many have failed to take over the world before. But things have changed.

All war has always been about one of two things. There is the physicality of war in which one side kills the other, brings about disease and suffering, uses fear and violence as leverage, and compels others to do their will. Total war, ala Clausewitz, seeks to destroy the enemy entire. With nothing left, the last army standing has won. But what have they won? Only survival?

The other thing that war is about, and has always been about, is the hearts and minds of the people. What good is it to be a dictator if your people don't love you? Eventually you can kill off those you find don't love you, but slavery is not really the desire of any dictator for very long. And no dictator can make everyone slaves. Someone has to be there to support the institutions of the winner, to continue to quell the public, to make the food and bathe the leader. And they, in the end, must end up trusted. So what is the basis for trust if not fear? It must be that the leadership has a believing public of adequate size and might to quell the opposition, for there will be opposition as long as there are non-believers.

In order to win a war, you must win the hearts and minds of the people that are left alive at the end of the war, no matter how few of them there are left. All of the violence is, in the end, only a tribute to the insanity of humanity. If you can win the hearts and minds, you have won without firing a shot, and if you cannot, you can never win while anyone is left alive.

Now I will admit that war still has violence and will for some time to come. But there is a world war underway right now and everyone in the world with a very few exceptions is involved in it every day. They just don't know it. And for many of those who are running the war, this is not only just fine, it is one of their goals.

My view, and I will try to back it up in this book, is that the world is now engaged in a global information war, that is has been for some time, and that it will be either forever, or until one of the sides that emerges comes out victorious. The war has very real casualties and we see them every day in the news. But most people don't see it the way I see it. They think that the people dying in Iraq are somehow engaged in a different conflict than the people put to death in jails in the United States, or the protesters killed or retrained in China, or the people who die of violence from hate crimes, or the children being exploited for all sorts of nefarious purposes, or the victims of kidnapping rings in the Caribbean and Europe, or the tribal murders in Africa, or the revolutions in South America. If I have left someone out, they have my apologies.

There are, of course, some more organized groups that engage intentionally in this information warfare, knowing exactly what they are doing, and doing so with malice. Most participants are merely pawns in the great game of chess that is information warfare. And many of the intentional warriors don't understand their parts or roles in the great play they undertake.

1.1 Overview

This book is about information warfare and the information war underway in the world today. There is a rich history of information warfare, but I am not a good enough historian to describe it well. Nonetheless I will use historical examples to the extent I am able to describe them, and hereby bait all real historians to write scandalous articles about how I have made errors in my historical recollections – just spell my name and the book's title correctly so your readers can judge for themselves.

It is my intent to cover information warfare in many different ways; including but not limited to perspectives on nation states, criminal organizations, religious groups, other sorts of groups, intensity levels, propaganda, politics, military applications, weaponry, targeting, legal issues, drug use, rights and privileges, wealth and power, and who knows what else. If these seem to be areas that are not logically connected, then you don't yet understand what information warfare is and is all about. So watch as I connect the dots.

1.2 What do I know about information warfare?

I know enough about how people think to know that introducing myself as an expert will greatly enhance the extent to which you will listen to what I have to say in this area. But I have also seen enough of these introductions that bring in pundits when we need experts that I have decided to give you the long story of how I come to the things I come to. If you are bored by all of this, please feel free to flip to a later page and take it for granted that I know of what I speak. Otherwise hold onto your hat while I tell you a bit about myself.

When I was young, I was interested in information technology. At an early age I built a mechanical computer, In grade school I used computers and know other children who were into computers and spent a fair amount of time with them learning about how things worked at a place called Project Solo at the University of

Pittsburgh. I was never actually in the thing, I just hung around with the folks there and messed with the computers. I learned the language PIL (Pitt Interpretive Language), programmed computers for my father's laboratory at the University of Pittsburgh, and as a Boy Scout, ended up joining my older brother's group at Carnegie Mellon University where we used the IBM 360 programming in Lisp and APL. I was not a very good programmer, but I did understand a fair amount about it. In high school I programmed computers and helped others who used the PDP 8 computers that were available, at that time.

When I went to college at Carnegie-Mellon University, I fell in love with electrical engineering and started to develop hardware and software for computers. I worked as a computer operator in the early days of the ARPAnet where I ended up working night shifts and chatting with operators in Germany and elsewhere over the net in the early 1970s. I worked on research projects including the design and simulation of protocols for military packet switching networks and as a low-level flunky on the Hearsay speech understanding project while I went to school. I graduated by the skin of my teeth in 1977 with a B.S. in Electrical Engineering.

After taking some time to find out that I didn't want to work for Collins Radio, RCA, or anyone else, I ended up getting my Masters of Science in Information Science from the University of Pittsburgh (top of my class for what it matters) with a thesis on robotics that nobody in their right mind would ever want to see or try to read. Soon after that I went to the University of Southern California to pursue a Ph.D. in Electrical Engineering, with emphasis on Computer Engineering. Somewhere along the way I did any number of other things, like designing a digital and analog timed permutation lock as a class project along with Brian Walters (Hi Brian), learning how people broke into computers and facilities (no names please), pushing buggies, and finding the woman of my dreams (who I eventually married).

While innocently working on my dissertation, which included classes in computer security and cryptography, I came across a strange idea one day in class and soon thereafter invented the first "*computer virus*". For the next 5 years or so this dominated much of my time and effort as I created any number of different defenses against viruses, including but not limited to most of the ones in widespread use today. I was young and dumb and didn't patent anything because I figured I would be a professor and the publications would count for more than the patents anyway. I'm trying not to make that mistake again. Along the way to my degree I ended up interacting with any number of folks in government(s), various folks at large corporations, folks with less honorable intents, and other folks of all sorts. Some of them were interested in the potential for the use of computer viruses in warfare (like all of the government ones), in crime (the crooks and corporate defender types), and in uses they wouldn't tell me about (like the Chinese graduate student I later had). Eventually I managed to get the US Department of State, the CIA, the NSA, and who knows who else unhappy with me, got black balled from research funding because I thought and still think that reproducing programs can be useful, and generally had all of the experiences that any other good information warrior will experience in their youth.

After graduation, I was a professor for a bit, then I ran a small becoming medium business (250 employees was the largest it grew before I left), a consulting firm (which I started when I graduated from college and still run today), and somewhere along the way started working more closely on issues surrounding information warfare. In the early 1990s while working under contract for SAIC for the *Defense Information Systems Agency,* I wrote a paper titled "*Planning Considerations for Defensive Information Warfare - Information Assurance*" which first defined the term "*Information Assurance*" as it is now used in the US government and most of the world (sorry to those who had previously thought of the term "Information Surety" but that's what happens when you don't publish these things). Eventually I took that paper along with other work I was doing and turned out a book called "*Protection and Security on the Information Superhighway*" which was not very popular in 1995, after which the President's Commission on Critical Infrastructure Protection concluded that all of the things in the book were about right and started a massive effort to fix up the critical infrastructures of the United States that, along the way paid SAIC a lot of money, but me, not a thing.

As I languished writing papers on these subjects that ended up falling largely on deaf ears and went further into debt while growing a family. I gained ownership of the all.net Internet domain, vied against Internet-based evildoers, and eventually ended up getting hired and moved out to California and work for Sandia National Laboratories. The things that closed the deal were that (1) when I left the interview, it was in the mid 60's in California and when I got back to Hudson, Ohio it was about 40 below zero, and (2) they agreed to allow me to do outside consulting as part of a condition of accepting the job, so long as it did not create a conflict of interest.

During my 8 years give or take at Sandia, I worked on a wide range of issues. This included but was not limited to critical infrastructure protection, research in information warfare, defensive information warfare technology development, research in deception for protection, digital forensics, cryptanalysis and cryptographic system attacks, the mosaic problem, and network centric warfare. I eventually started the Sandia College Cyber Defenders program with lots of help from others who tilted at windmills, like Dick Isler, Fran Dreshler, and Nina Berry. This program was the precursor of the US cybercorps that is now used to train information security professionals across the US. I trained red teams for investigating system vulnerabilities, worked before the end of 1999 to help assure continuity of the electric power grid, helped design and implement systems for classified information processing, did research in deception for information protection, and implemented some defensive information warfare systems now in use within the US department of defense.

At some point, I started teaching graduate classes at the University of New Haven as a sideline, and of course I continued to operate all.net and do various other information security related work like the development of the Deception Toolkit and the White Glove Bootable Linux CD, and worked on occasional investigations and forensics matters in my dubious spare time. I also started the information warfare mailing list in the late 1990s, an Internet-based list for serious discussions of issues in information warfare.

At the University of New Haven, I started teaching a set of courses in subjects related to information protection, including but not limited to secure network infrastructure design, cryptography, computer viruses and related attacks and defenses, and eventually a class in deception for information protection, a research course in investigating cyber terrorism, and other similar classes. At the University we did some intelligence-related work for elements of the US government, taught courses to people working for agencies like the FBI, military intelligence services, and a wide array of other in service and recent graduate students. I also taught courses in digital forensics as a California POST certified instructor, helped develop national standards for digital crime scene investigation, and participated in the New York Electronic Crimes Task Force.

The University, or more specifically, Tom Johnson and John Tippit, had been working on starting the Masters of Science in National Security program for some time. This program, unlike other programs in national security around the world, was focused on the technology, science, and governmental structures that form the national security apparatus, and not on international relations or how to get along with the world. It had all of the necessary approvals to start the program in the fall of 2001, but the plan was to start on January 2, 2002 with the first class. After the attacks of 9/11/2001, the program started up, but despite appearances, the startup had nothing to do with these attacks. The attacks did help to focus students on the program.

Just before the University started their program, Sandia fired me for reasons that were and remain a bit murky, but I think it's because I offended a director one too many times. For some reason, they seemed to think that losing my Q clearance at termination was a big deal, but it never did me any particular good when I had it, since every classified thing I ever heard about in my work was something I already knew from the unclassified world before I heard it was classified. The only real utility of telling me any of it was to keep me from talking about it in public anymore. I went to meeting after meeting where I ended up asking what was classified about the content of the meeting. They would say something like “the whole thing”. And I would start to go through item after item indicating where it was published in the open literature. After a few hems and haws, they would tell everyone present that the thing that was classified was that they had said it. So I have made certain not to tell you who had these meetings, even though the meetings were announced in open emails at the time.

I continued doing consulting for a bit before being hired by Burton Group as an analyst doing “*research and analysis*” of the security and risk management strategies space and industry. This is a job with incredible access to information and people, and it brings enormous clarity in respect to what motivates different people and groups who form the societies we live and work in. On a regular basis, analysts will talk to top executives and top technical people in most of the largest and many of the smaller vendors that sell into the security space, many of the top decision makers in the large enterprises that operate across national boundaries and control and influence many billions of dollars each, many of the top technical people in the field who work for these enterprises, and a wide range of others who are involved in some other fashion. We get invited to all manner of parties and meals, essentially all of which I end up turning down because I want to remain as independent as I can in evaluating this space. Our team also ends up writing reports on parts of the security space that help to define how the decision makers at these large enterprises think about the issues and make their decisions.

For now I also:

- Own and operate *Fred Cohen & Associates*, which does select consulting work, gives talks at conferences, runs *ASP Press* that published this book, runs *Totful Toys*, that sells games and high-end playful dazzlements and related technologies, and provides an outlet for me to do and say whatever I like without worrying about who is offended.
- Oversee and chair *SecurityPosture*, Inc. out of Omaha, NE, which does wireless networking implementations and security, supports secure infrastructures for enterprises, and does other sorts of security-related consulting.
- Teach graduate courses at the *University of New Haven* over the Internet, telephone calls, and sometimes in class.
- Work with Tom Johnson helping to form and forward the purposes of the *National Security College* which will, someday soon we hope, be teaching courses in affiliation with your local institution of higher learning.

Add to this the time I spend time with my family, our dog and cat, finding ways to pay off the mortgage every month, and my recent foray into SCUBA diving, and you get the idea both of the life I live and the extent to which I think about the issues of information warfare. With this in mind, and having fulfilled the mandate to present the reason I am qualified to be considered in terms of the issues of this book, I welcome your indulgence, attention, and comments on the substance and my views of it.

2 What is iwar and why it is important?

Information warfare (iwar) has been studied in various forms for a very long time, and yet the definitions of it are still varied across a wide range. I start with a definition that I think has some merit:

> *Information is symbolic representation in the most general sense. Warfare is high-intensity conflict between opposing parties. Information warfare is about manipulating and protecting the symbolic representations used and targeted in high-intensity conflicts.*

Now this probably seems like a really strange definition to most readers who are not familiar with the field, and somewhat less strange to those within the field. I will explain it by discussing other views and relating them.

2.1 Network-centric warfare

The most common perception in the public of information warfare is what some in the US military came at some point to call *network centric warfare*. This is where warriors use computer systems and networks to attack opposition computer systems and networks and try to keep their opponents from doing the same to them.

It is important to note the common threads. There is always attack and defense in warfare, and there are always at least two sides (them and us). Network centric warfare focuses on computers and the content they bring to use. It is the utility of the content that is ultimately at issue here.

There are always:

- Attack
- Defense

There are always at least two sides:

- Us
- Them

2.2 Objectives

For example, if the attack is intended to disrupt network operations, the real goal is to deny the enemy the utility they would normally have from the content, and perhaps to consume their resources in trying to regain that utility. If the attack is corruptive in nature, the idea might be to alter the utility of the content to favor us. If the goal is to leak information, then we are trying to gain the utility that access grants us and perhaps reduce the utility of the same information to them. An attack designed to defeat accountability is typically used to manipulate the results of applying content so as to gain a financial or power advantage without them knowing who did it. And an attack designed to defeat use control grants us the ability to use their content to gain utility against them and keep them from using it to gain utility against us. On the defensive side, we are trying to prevent them from doing all of these things to us.

Objectives include either protecting or defeating:

- Integrity
- Availability
- Confidentiality
- Use control
- Accountability

In order to understand this more clearly and deeply we need to understand how content is used in warfare; What is its utility? This of course depends a lot on the specific them and us involved. The US military, and most military organizations in the world, use content in an enormous variety of ways. For example, automated weapons systems like missiles use content to determine where the missile goes. Thinking in terms of the defensive objectives of integrity, availability, confidentiality, use control, and accountability, failure to meet those objectives could result in retargeting the weapon against our own troops, making the weapon fail to arm when deployed, alerting the targets that the weapon is aimed at prior to deployment, causing the weapon to explode just before deployment, or being able to take and sell the weapon to them or others without getting caught.

2.3 Mismatches

Definitions are very important to information warfare because the way people think about the issues drives the allocation of resources and the focus of attention they place on different things. Because winning battles and wars is very often about creating mismatches and because military organizations tend to be hierarchical, a poor or improperly matched definition by a high ranking individual can drive a military down the path to defeat. A good definition can, of course, bring strategic advantage and victory after victory.

Here is a definition from a US Air Force General directed at intelligence gathering, information technologies, and self-defense, in that order. I think it shows a certain level of confusion, but I also think that this is part of the process of finding good definitions that the military organizations of the world go through in their quest to deal with changing times:

> *"First, IW includes those actions we take to gain and exploit information on the enemy. Second, IW includes what we do to deny, to corrupt, or to destroy our adversary's information databases. Third, how we protect our systems must be included as part of IW."* USAF Chief of Staff - General Ronald L. Fogleman

2.4 PsyOps

> Psychological operations are about protecting our mindsets and manipulating theirs.

As we move away from network centric warfare, there is no need to move away from the same basic principles I have been discussing. For example, let's look at psychological operations, also known as PsyOps. PsyOps is about using results of psychological research as operational tools. The targets in this case are the minds of people. Thinking in terms of the framework for information war we are building, PsyOps is about protecting our mindset and manipulating their mindset. For example, we might want to keep ourselves confident and make them afraid. Of course the most successful outcome of all would be to get them to have the same mindset that we have because then they would be on our side and the conflict would be resolved. But it turns out not to be so easy to get people to change their minds.

2.5 Them and us?

Of course there are other dimensions to consider, and consider them I will. Suppose we drop the notion about them and us for a bit and consider that, in reality, there are a lot of individuals involved in conflicts, and not all of them are on one side or the other. We may start to see that the "*them and us*" perspective is not only naive, it is problematic. The whole notion of *them and us* is flawed in reality, but from a practical standpoint, how do you get your warfighters to kill some other mother's child unless you depersonalize the acts of violence? One of the lessons of PsyOps is that you need to indoctrinate your own people in order to defend against enemy PsyOps and keep your people from killing themselves and others over the tremendous guilt that killing brings to some people. A selection process that identifies people who are unsuited to this process would be most helpful, but nobody has a good one.

"*The enemy of my enemy is my friend*"

"*With friends like that, who needs enemies?*"

In the world as it is today, there exist interdependencies between societies that are both enormous and unprecedented in the history of the World. If conflict turns into a situation where "*you are either with us or against us*", duplicity will result. Plenty of people will tell you they are with you and they may even act a little bit like it to keep from being treated as if they are against you. But in reality they will pull you apart from inside. Alliances are typically based on interests, and interests vary with circumstance. If you are winning, they will smile and stay on your side, but if you are losing, they will smile at the other side and support them in subtle ways. If the leaders who are responsible for the war don't understand this and find ways to control the situation, it will spin out of control all on its own.

2.6 The spectrum of conflict

I will be talking about intensity issues soon and in more depth, but as a basic principle, it is important to understand that not all warfare is totally destructive. More generally, conflict ranges over a broad spectrum from almost completely cooperative and friendly with the most minor of disagreements to the sort of rage seen in hand to hand combat and the extreme violence of nuclear weapons. All war is not total war, and information warfare in one form or another exists at all levels of intensity.

Conflict also tends to wax and wane with time. People have only so much energy. They can get enraged, but they long for peace, and many people can't stand to live in peace and calm all of the time and seek adventure and excitement. Societies become anxious for conflict when properly prepared, but they tire of them over time, they exhaust resources, tire their fighters, create enormous burdens on the society, and wear down resolve.

> "*if the campaign is protracted, the resources of the State will not be equal to the strain... There is no instance of a country having benefited from prolonged warfare.*" -Sun Tzu, *The Art of War,* 1910 translation of 5,000 year old the ancient texts.

2.7 Certainty and intelligence

Many view information warfare as inextricably tied to intelligence, and certainly intelligence in warfare is about gathering useful content about the enemy. But countering the enemy's intelligence efforts is also a critical element of information warfare. From an offensive standpoint, the goal is to gather, fuse, analyze, and evaluate information so as to increase your certainty of the realities you face, both about the enemy and about yourself. In addition, I will call it offense to decrease the certainty with which the enemy knows the realities about you, however, this is a great simplification in that there are certain realities that are intended to be projected toward the enemy. On the defensive side, the goals are to prevent the enemy from decreasing your certainty about the reality and to prevent the enemy from increasing their certainty. If this sounds like Donald Rumsfeld discussing how we know what we know and don't know what we don't know, you have it right. A phrase to keep in mind is "*the fog of war*".

Offense:
- Increase your certainty
- Decrease theirs

Defense:
- Retain your certainty
- Don't let them keep theirs

2.8 Tempo and time

Information warfare has been described by many in terms of the impact of information technology on warfare. This comes in several major areas from an offensive standpoint. One of the most important areas is the implications for time and the tempo of operations. Tempo is the rate at which things can be done, and as Boyd pointed out in his work on the Boyd cycle, the rate of the decision cycle along with its accuracy determine to a large extent who wins and who loses battles. If you can observe, orient, decide, and act faster than the enemy and do it with the same or greater precision and accuracy, you will win almost every time, and by a great margin.

The Boyd Cycle:

- Observe
- Orient
- Decide
- Act

Nowhere was this more clearly demonstrated than in the fist Gulf War in a particular battle called "*The Battle of 73 Easting*". In this particular battle, several US tanks came up over a small berm, and as they emerged they encountered scores of Iraqi tanks, all loaded, fully manned, and ready to fight. Over the following minutes, these US tanks killed every one of the enemy tanks and did not suffer even one death by enemy fire. They did it because they had faster tempo. They were able to observe the situation, orient themselves to it, make decisions about what to do, and act before the Iraqi tank commanders could target them. And they were able to keep moving while firing accurately at target after target. This was the direct result of the use of information and information technology in their tanks and of their training in how to do battle at this pace. The ability to act faster and more accurately is an advantage brought about by information technology that is so great that ten to one odds are no problem to overcome with a significant tempo advantage. Looking at relative casualty counts, the advantage in that war was even greater; on the order of more than 100 to one. Clearly, information war in these terms is fundamental to success.

2.9 Targeting

In describing the Boyd cycle, I also described another enormous advantage relative to the fog of war. It was not only the rate at which US warfighters could act that won this and many other battles. It was the ability to identify, locate, and hit targets with higher accuracy and more often that also won the day. And again this is a place where information technology has dramatically altered the nature of armed conflict.

> The ability to find and kill distant targets using complex infrastructures in real time is unparalleled in history.

The ability to target a particular weapon on a particular location where you know the objective lies without expending excess resources and while limiting collateral damage is also an incredible advantage, both in terms of efficiency, and in terms of perception. This is largely the result of advancements in information technology. The current situation is incredibly complex, but a simplified example will do to clarify.

Suppose I want to find and kill enemy weapons that might be interfering with my plan to take the next hill. The process involves data collection in the form of everything from ground troops in hiding to satellite imagery in real-time. From this data, there is a fusion up and across different echelons until the lowest echelon that can have access to all of the data required to identify and locate targets. Targets may be identified and located by people and their systems at a location far distant from my group, and the fused data is presented on a display that shows me everything within a few miles of my position. I then select the targets of choice and make a request for weapons systems located miles away to put weapons on these targets within the next few minutes. Those systems take the targeting information, send off their weapons, and coordinate the activity across hundreds of different similar simultaneous activities, landing the right weapons on the right targets so that when I go over the next hill, I will face little or no resistance.

2.10 Interdependencies and brittleness

All of this complicated stuff that has to happen in order for this warfighting result to take place means that there are a lot of opportunities for failure. If the signals detected are wrong, the analysis incorrect, the fusion corrupted, the presentation in the wrong color, if any element of the communications or computation is unavailable, if the targeting is in error in any way, if the missiles have an error, or if the enemy finds out what is happening, the game is up and the overall system fails to accomplish its mission. The result is that the soldiers going over the top of the hill meet strong resistance and there are dead and wounded on both sides instead of just on their side, or even worse, all of the casualties on our side of the hill and the enemy occupying it.

This returns to the previous harkenning to integrity, availability, confidentiality, accountability, and use control. These protection objectives are absolutely central to winning the information war. But this example is of course highly limited compared to all of the elements requiring effective information for success in war. Supply and logistics, battle damage assessment, procurement, troop deployments, strategic decisions, tactical decisions, everything in modern and historic military activity depends critically on the content and its proper use.

> The same infrastructures that support the military, support our whole society... This makes them legitimate military targets.

Lest you come to believe that this is only a military issue, consider that the information infrastructures that support military operations are integrated at every level with the infrastructure elements that all members of modern society depend on for their survival. The same power supply that supports military communications supplies civilian populations. The same information infrastructures support both military and civilian communications. The same supply and logistics chains form the back end of both military and civilian societies. We stand together or fall together. These infrastructures are legitimate military targets.

When civilian infrastructure becomes a legitimate military target, civilians suffer in the eventual wars. Protecting these infrastructures implies adequate redundancy to assure availability, adequate protection and operational controls to protect integrity, widespread secrecy techniques to assure confidentiality, use control that ultimately demands identity association between large numbers of civilian populations and what they are, have, and know, and accountability ultimately means surveillance on an unprecedented scale. And that is exactly what we see in the US today.

For those who lived through the second World War, these things are abhorrent because they were the very things used to quell civilian dissent. And of course they are used in China and Indonesia, and elsewhere today to quell dissent against autocratic governments. Is that what the US is headed for as well? This is certainly an information war issue, but it is information war against the population that leads to the potential for military revolt or emergent dictatorship, even if presented as if it were democracy.

2.11 Mapping and personal warfare

> The use of mapping at the level of granularity now available leads to personalized threats and home targeting by military or government officials.

Another area where information technology has integrated war capabilities with civilian capabilities is in the area of mapping. All you have to do is look at the Google.com entry into the mapping and presentation of satellite imagery and you will soon come to understand the level to which a government official or member of the military could direct death and destruction against individuals. Suppose you are doing something I don't like or opposing me politically, or just accurately reporting the facts and questioning my propaganda. How effective is it to send you a message in the form of a picture of your house from above, your address, names of your children, where you work, your financial situation, and so forth. The threat is very simple and very real. Stop it or I will kill you and your family at the moment of my choice with fire from the sky.

That is exactly the threat carried out against Muammar Qadhafi when his compound was bombed killing his family members. The capacity to deliver on targets and map out where somebody lives has increased so much over the last 20 years that anybody with minimal skill in electronics and access to explosives and a model aircraft can potentially make a weapon that can target an individual home and deploy explosives into it from 25 miles away. The home made cruise missile is within the grasp of individuals today and certainly an easy development for any small government or substantial militant group. The capacity for individualized war is indeed an enormous breakthrough.

2.12 Hard kill vs. soft kill

In some cases it is not necessary to kill anyone in order to effectively disable their ability to resist. Disabling their capability to observe, orient, decide, and act may make them as good as dead from the standpoint of defeating their utility. Consider the soft kill approach taken in political elections and propaganda campaigns. The marginalization of individuals based on false rumors combined with reducing or eliminating their capacity to use the media to put out their message marginalizes their effectiveness and eliminates their capacity, and sometimes even their will to fight.

Soft kill retains the capability after the battle. This reduces the cost of winning the peace.

In more direct military use, it is often desirable to prevent an enemy from using their capability without destroying the capability for future use. For example, the use of disabling carbon fibers across power lines denies enemy use while allowing rapid repair for your use after the battle. The cost of rebuilding enemy infrastructure to win the peace is dramatically reduced with soft kill instead of hard kill weapons and tactics. In information technology this is particularly easy to do. Disabling a computer system may be quite simple while the content remains completely intact for future use. Do this in a control system for a critical infrastructure and the enemy capability is disrupted temporarily and repaired in a matter of a few hours afterwards.

2.13 Economic war

If winning the war involves swaying the hearts and minds of the enemy, the Cold War was an example of winning the war without firing a shot. Some shots were fired in the cold war, but for the most part, it was a war with no battles. It was fought with pure strategy because neither side was willing to assure its own destruction by attacking the other directly. In the end, it was an economic war, not a nuclear war, and the Soviet Union literally lost its capacity to fight as it lost its ability to sustain itself. The US is having a similar conflict with China and is having problems sustaining itself as did the Soviet Union. Sun Tzu had it right when he said:

> *if the campaign is protracted, the resources of the State will not be equal to the strain... when your weapons are dulled, your ardor damped, your strength exhausted and your treasure spent, other chieftains will spring up to take advantage of your extremity. Then no man, however wise, will be able to avert the consequences that must ensue.*

> In the information age, money is represented by bits. A war can quite literally be won and an economy destroyed by network centric attack on the financial system.

While most wars are about economics in one way or another, in the information age, as information has literally replaced other fungible financial instruments, that information is subject to direct attack in the form of network centric warfare. National economies can be ruined by successful attack on computer systems. Move all of the balances in all of the accounts so that transactions fail, the rich become poor, the poor become less poor, and a few people here and there end up with the representations of wealth. The tangle gets so deep that nobody can undo it. Ask the Japanese company that recently lost almost 300 million dollars in a day when an error in a computer entry that could not be repaired by the Japanese stock exchange in time resulted in enormous numbers of shares being sold for 1 Yen each - almost a millionth of the desired offered value.

2.14 Controllability of effects

In most uses, weaponry is more useful if it can be better controlled. While simple explosives have limited radius of effect, it is far harder to control the secondary explosions in an oil refinery once the first tank of gasoline is blown up. Information tends to have transitive effects. For example, computer viruses are commonly designed to take advantage of the disease-like spread of infection to go from place to place seeking out all of the available paths and extending itself transitively. But just because this can be done doesn't mean it has to be done. A far more effective approach comes from the ability to demonstrate control over effects.

Consider the desire that could be created in the enemy leadership to stop fighting if they have it clearly demonstrated to them that you can kill anyone in their family at any time, take all of their money away, create the false impression that they are misleading the country into horrible outcomes, make them look ridiculous, and disrupt their control over their people. You show them that you can several times, then explain what you want of them. If it is reasonable, they will usually comply.

> Ultimately, demonstrating clear control is a compelling reason for the enemy to yield. As your control gets tighter, it becomes harder and harder to deny you what you want. But control is not the same as agreement.

Controllability is good, but reversibility is even better. Suppose you take their money, disrupt their power base, destroy their support, and kidnap their children, then have the capacity to reverse it all. While this is not likely to work with most war activities involving violence, when effects are less destructive, reversibility can be played to demonstrate such clear control that it is hard to fight against it. The more you can control all of the things I have been discussing, the better off you are in winning the war with minimum negative consequences. But control will not win the hearts and minds of those you control. It is more likely to bring out resentment that is suppressed from lack of choice.

2.15 Control of the media and the message

As control grows on a larger social scale, control of the media becomes feasible. Of course there are many societies in which the media is owned and controlled by the government. China is this way, so is Singapore, and then of course we have the US where control is so great that everybody thinks it is uncontrolled. I will go into this later and in more depth, but for now, the focus will be on the more direct sort of control demonstrated in so many wars through the pamphlets and public relations efforts of the military organizations, and its support by control over the communications systems and media.

When the Soviet Union was near its end, there were struggles from time to time involving attempted takeovers. And in every case, the fiercest fighting took place for control over the state owned radio and television broadcast studios. The game was a simple one indeed. If you take over the media, you can give the appearance that you are in charge. Do it for a minute and get killed and you are a nut case who got what they deserved. Do it for a few hours and you are the leader of a failed coup. Do it for a week and you are the leader of the nation. You start ordering the movement of troops and preventing other command and control from working. You get the people to act in different ways and grow support on a massive scale. An audience grows and hundreds of thousands of people show up at your behest to protect you and the station from the military, many of whom are now on your side, And soon the previous government officials are in exile.

Dulles took about 150 men and, with support of radios and some bombing runs managed to take over Guatemala without killing anyone.

You don't have to believe me about this. Look at what Dulles did in Guatemala. In terms of information warfare, few examples can ever compare with this one. With about 150 troops, he created a campaign that took control over the country. Details are well described at:

http://www.swl.net/patepluma/central/guatemala/vozlib.html

2.16 Truth and Lies: Deception in warfare

Deception is fundamental to warfare for without deception, the enemy would always have high certainty in everything they knew about you and you would have high certainty in everything you knew about them. In addition to the fact that this would not be very much fun, the fog of war has always been part of war, and in fact it is usually the deciding factor in war.

> Deception carries great risks and great rewards. Skill in its use often separates winners from losers.

Knowledge leads to the ability to reveal or conceal potential mismatches that can be exploited to advantage. Success in war comes largely from creating and exploiting mismatches, while failure often results when mismatches are exploited against you. Advantageous mismatches are sought. Disadvantageous ones are avoided. With the use of deception, you can end up winning battle after battle despite less of everything else, you can break the will of the enemy to fight, you can summon additional resources and create alliances, and so forth. And you can even lose all of the battles and still win the war by directing the perception of the enemy so as to lose the will to fight.

Deception revealed leads to serious credibility problems and even more serious battlefield problems. If deception is used to create a mismatch and the deception is revealed to the enemy in time, the results can be disastrous. Suppose a small force is being used to distract the enemy as a larger one repositions. If the enemy understands this, not only will the smaller force be cut off and made ineffective or destroyed, the larger force will be attacked at its weakest position and when it is least prepared to defend itself.

But while deception in battle can produce enormous losses, deception in the perception of populations involved in supporting wars outstrips battle effects short of weapons of mass destruction. The will of the people, the hearts and minds of the populace, is the place that deception is absolutely fundamental.

2.17 Propaganda

The hearts and minds war is usually won or lost in the propaganda efforts. Propaganda runs over a continuum of intensity just as the use of force does.

Generally, propaganda is considered bad when they do it and not as bad when we do it. After all, we do it to help keep our population willing to support the critical war effort needed to defend ourselves and to break the will of the enemy and stop their unjust and immoral aggression. They, of course, use propaganda to incite their people to violence and overthrow everything we know is good and right by manipulating the innocent victims of their domination, turning their population into evil mindless slaves to the state, while they use propaganda against us to seed lies and mistrust into our innocent citizenry causing them to no longer support our own children who are fighting to maintain our way of life.

> Propaganda is almost entirely a matter of perspective. Not!!!

Of course this statement is true for all sides. As ridiculous as it is that they would tell their people these scurrilous lies to justify the acts of violence and aggression that they have made, we say these same things to our people. Of course, when they say it, it is propaganda and evil, but when we say it, it's the truth that we must get out to our people in order to keep them from being lied to by the enemy.

The inability to separate propaganda from reality is fundamental to the success of propaganda and one of the main reasons that secrecy is so dominant in military propaganda campaigns. It ranges from changing the name of the War Department to the Department of Defense, to calling preemptive war self defense, to excusing the surveillance of your own population, to the intentional destruction of your political enemies. And if you think I am talking about the US, only one of the examples here (changing the name) could not be associated with Nazi Germany. They fought Poland in self defense, spied on their own people, and destroyed political enemies.

2.18 Marketing and operations

I have discussed more rapid Boyd cycles, lower cost, more selective targeted approaches, managing perception, soft vs. hard approaches, truth and lies, messaging, and the interaction of people and technology. I hope it was not lost on my readers that all of this sounds more like marketing than like fighting. Indeed, it does. And indeed many in business now apply the lessons provided by Sun Tzu and other warrior philosophers to gain advantage in the competitive world of business.

> Marketing is low intensity information warfare. As business gets more and more competitive, the intensity of the conflict increases and marketing gets to look more and more like war.

The reason businesses do this is because they are in low intensity conflicts. The intensity of conflict is on a continuum. The whole big push toward increased business competitiveness means that people feel increasingly threatened and more aggressive people are more successful when the intensity of the conflict increases. People start to think they can gain an advantage by doing something that is socially marginal, or if they think they won't get caught, or if they just don't care about you as much as they care about themselves, or if they live in a society where the norms are of higher intensity conflict. They will do increasingly severe, aggressive, deceptive, and intense things to compete in their environment.

In the US today, we hear about the notions of a civil society and we also hear about increased competitiveness. We see cost driving business efficiency, and we see shorter and shorter time frames for higher and higher performance. The more and further we go in this direction, the harder and harder it will be to keep up, and the fewer and fewer people will be able to share in the dream of success by these standards. For some reason, the people who talk about these things don't understand that they are in competition with each other. A civil society that is intensely competitive and yet completely moral is at odds with the nature of human frailty.

2.19 The free press will save us

Of course many in society think that it is the job of the press, the so called fourth estate, to save us from the propaganda and marketing operations of government and business. After all, the press is free in most Western civilizations, isn't it?

Sorry to dispel yet another rumor. The press is not free. It can be relatively low cost if you can spend the time and effort, but for real success, it is downright expensive.

> Getting the "free press" to tell you all about this book cost me something like US$5,000. Imagine how many I could have sold if I had more money to throw at the free press?

Freedom of the press means that the press is allowed to say more or less whatever they want to say. And this is more or less true in the notional sense in most Western societies. But just because they can theoretically say whatever they want, doesn't make them free in the sense that you don't have to pay for it.

The cost today of sending a press release to about 40,000 journalists in the US is on the order of $500. This is very low cost indeed and it leverages the Internet for great efficiency through the use of email as the delivery channel. It used to cost almost a dollar each, now it's a penny each. Great bargain. If you want to facilitate a relationship with the radio media. For about $1500 per national interview show and $200 per local radio show, you can schedule slot after slot during drive time when the most people listen to the radio. To sell you this book, I will have paid something like $5,000 in "Free Press" costs so that the free press finds out about the book and tells you about it. My last book involved more like $10,000 in "free press" costs. Why is this?

Because our friends in the media don't search out the news, it is spoon fed to most of them through public relations efforts. Some seek it out, but mostly, what you see in the news is propaganda by another name. Public relations, marketing, and paid advertisement.

2.20 The social fabric and dependencies

As you and I break down the issues in information warfare, it becomes increasingly apparent that it is being used against us.

When the free press costs so much money and one mother's child is worth more than another because one mother can interest the press while the others cannot, the social fabric that makes civil society work starts to break down. How many poor black mothers are in the media for months at a time when their child is kidnapped, raped, and killed? None!

When those supporting a war are all over the media, and those who oppose it are shunted to the side, don't get “air time”, and are followed and spied upon by the military within their own country, and some of them are arrested on conspiracy charges or carried away without access to a lawyer or any notification of their family, and when religious leaders openly endorse political candidates who send money to their churches, the world around you is engaged in an information war.

> *Again I want to remind those in the US who may read this as criticism of the current administration that I am talking about Japan just prior to Pearl Harbor, or was it Iran just prior to the war with Iraq, or was it Germany in World War II, or was it China just prior to Tienanmen Square, or was it Venezuela just prior to the revolution (each of them).*

National sovereignty, supply and logistics, military sustainability, national will, moral, and so many other things are critically dependent on the information infrastructures of a country and the World, both because of the direct effects that failures or corruption of those information infrastructures have and because, as the supporting elements start to crumble, the very fabric of societies unravels. The Soviet Union came undone because of financial and social collapse, not because of shots fired or bombs dropped. Societies exist because they create a fabric of interaction and commerce between people and they collapse when the fabric fails.

2.21 Reality eventually sets in

No matter how much the government or the media keeps telling us that perception is reality, it is not. Reality eventually sinks in and things like how much food, housing, heat, electricity, and gasoline cost out of the real paycheck people get hits home. The excuses can keep coming, but eventually, revolutions result. The "*let them eat cake*" view of the world is typically one embraced by leaders out of touch with the reality on the ground.

While there is a limit to what people will believe in the face of facts to the contrary, that limit may be further than most people believe. Sacrifice for the war effort, others are doing well but prosperity has not reached you, it was the weather not the government, we are a poor people and it is all we can do to survive, the enemy is systematically trying to starve us, they are putting too much pressure and we can not, we will not stand for it! These all work... for a while. But then...

> When reality sets in, woe be to those who have pushed deceptions and perception management too far. A bloody revolution often follows gross overextension of information warfare against the people.

Reality sets in. Trust erodes in a government that says one thing and does another. If they say it is better elsewhere and prosperity hasn't reached you yet, after a few years, you may conclude that it won't reach you and that the people telling you it will are just plain lying. After they misname one program after another to indicate exactly the opposite of the reality, the names become symbols of hatred and of the lies perpetrated. People today watch the media all over the world, and that means competing views. China tries to stop it by controlling the media, as do others, but it is a very tough row to hoe indeed. Over time, projections of realities from elsewhere come out. Images from around the World are available to most of the World, and the suffering of people after natural disasters and the conditions that were apparent prior to those disasters are obvious.

2.22 A long way to go

Clearly the issues of information warfare and, more generally, conflict in the information age, are very complex and cover a broad scope. And just as clearly we have a long way to go in resolving these issues and a short time to get there before we lose the ongoing World War – or should I say global conflict.

But before I go on to the more detailed coverage of this book, I think it is worth a short pause to ask a simple question:

Who are we?

I don't mean to ask the complicated question about the nature of life and everything. That is the sort of thing that folks like Douglas Adams asked and answered so enjoyably. I am no hitchhiker in need of a guide to the galaxy. I mean to ask and answer the really simple question. In this book, I talk about us and them a lot. I just wanted to reiterate who we are when I talk about us and them.

You and I are we. We stand together or fall separately.

In the information age, it's everyone for themselves. However, the best way to be for yourself is to cooperate with others. My point is, that whoever you are, you should consider we to include me. After all, I don't want you to use global mapping, a cheap GPS system, a model aircraft, and some Internet instructed explosives to target my house. Just remember, I am a poor, innocent, hard working writer who is just trying to help us and not the evildoers on the other side.

If everyday people of the world don't stand up for themselves, the rich and powerful will take over. It has always been this way and the history of humankind convinces me that this will continue. Information warfare must be practiced by every person in the information age or they and their way of life will perish. It is survival of the fittest one on one with the world. It is the ultimate World War, and every person in the World is individually at war or in alliance with everyone else. Who will win the peace? We will. You and I.

3 Intensity levels of information war

Information warfare, as all warfare, varies in intensity over time and by situation and location. Low intensity warfare is often associated with political disputes that get out of hand, protests, and perhaps even riots. Different sorts of things come into play as the intensity ratchets up and they are put back away as the intensity ratchets back down. In addition, the lessons of high intensity warfare are of enormous value in understanding and winning lower intensity conflicts. Indeed, if we get good enough at dealing with these issues, there may not have to be high intensity conflict any more.

Ultimately information has to be a fundamental basis and cause for all intentional conflict because some thought process has to be involved in any intentional act. Of course we could end all human wars by simply altering the brain chemistry of all humans, disabling the amygdala, removing the frontal lobes, or other similar acts of mutilation.

> The reason to study intensity issues is that there is hope that conflict may be reduced in intensity and thus reduced in terms of human suffering that ultimately results.

But the hope of peace for all humanity is not that all conflict ends. That would also end the competition that is the reason people grow and humanity makes progress. Progress can be measured, if only in terms of increasing human life span, decreasing pain, increasing qualities of life that we think to be important, and decreasing disease and debilitating injuries. Rather, the hope is that the occasions in which the level of intensity goes beyond what we expect in a civil society are rare, and disputes are settled in more humane ways that don't involve killing large numbers of people who had nothing to do with the conflict but were dragged into it by those who wished to make war.

3.1 Offense

Generally, conflict involves offense and defense, and certainly conflict involving symbols has both offensive and defensive elements to it. In most cases, the number of offensive options are larger than defensive options and the nature of offense is that it is better funded, treated as more important, and emphasized in the study of war. The old saying goes: "t*he best defense is a good offense.*" It is true in football, unless you watch teams from the North Eastern part of the US on snow days, and certainly you can only hope to not be defeated with only a defense.

In information offense implies creating information aimed at the enemy, using force to reduce the enemy's information capabilities, the use of tempo and targeting against the enemy efficiently, and all of the other things discussed earlier. The thing that makes these offense instead of defense is that we are taking the action to them. We create opportunities, we exploit those opportunities, and we control the agenda and the action.

3.2 Defense

Defense uses some of the same tools, but acts to assure that the information capabilities we have are sustained at a usable level during the conflict. While some may associate this with reaction instead of proactive approaches, this is a fallacy. Defense takes as much if not more planning and preparation as offense, but it tends to produce fewer dramatic results. In fact, the whole goal is to eliminate the drama. A sound defense will result in no surprises, everything working as it should, and the enemy never getting close to causing any problems.

> In the information arena, the best defense is not a good offense, It is a good defense. Attackers tend to make poor defenders.

To dispel an old fallacy about defense, the best attackers in the world do not make very good defenders in the information arena. And no amount of attack will reduce the need to defend your information assets.

3.3 Overt, high frequency, tactical actions

At the high extreme of one dimension of the standard two dimensional conflict chart, we have overt, high frequency, tactical actions. They tend to be overt largely because they are highly aggressive and rapid, aimed at moment-to-moment issues of gaining an advantage over the adversary.

These include things like a media blitz by the President of the US, or the shock and awe campaign against Iraq, or the dropping of leaflets into a country as part of the psychological operations in a war.

> One dimension of the conflict spectrum relates to the extent to which the enemy is aware, the rate at which things happen, and the time frames and level of decision-making involved. This is typically presented on the vertical of a conflict diagram representing the escalation and deescalation processes of the conflict.

3.4 Covert, low frequency, strategic actions

At the low extreme of that same dimension are covert operations that are lower in frequency, and oriented toward gaining a strategic advantage. They tend to be covert because they are aimed at longer term goals and, if too obvious, can be countered by a wise opponent. They tend to be low frequency because if they get to be too frequent, they become rather obvious if someone is bothering to watch for them. They are strategic because they are aimed at a permanent advantage, typically one great enough to warrant the high investment over a long period of time that they take, and they tend to be oriented at top-level enemy decision making processes relating to overall direction that takes a lot of effort to change.

These include things like the planting of Trojan horse programs or hardware in enemy systems, elicitation approaches at gaining intelligence, or the creation of strategic deceptions oriented toward altering the investment approaches of the enemy.

3.5 Peace, competition, and diplomacy

At the low end of the conflict spectrum we have peace with a normal level of competition associated with the human condition, and ongoing diplomatic relationships with other parties. This is business as usual for most of the people in the Western world most of the time since World War 2.

Of course there have been plenty of wars and conflicts along the way, and the military industrial complex we were all warned of at the end of World War 2 has been working to sell weapons to the World and various folks here and there have fomented wars of all levels of violence along the way. But still, most people in most of Western civilization have been working in industrious ways and making progress toward raising all boats along with their own over that period.

> Peace, normal competition, and diplomacy are the best condition for the people of the World. But unfortunately, both natural and human forces are often present and they move the world away from peaceful coexistence.

When conflicts arise, diplomatic solutions are sought. For example, when trade barriers are put up because of poor economic conditions in one place or another, diplomatic solutions are sought, trade bodies are created to settle disputes, and even when there are sanctions of one sort or another, they tend to be fairly restrained. This works to the advantage of all since it prevents those trade wars turning into violent conflicts where lots of people get killed.

Of course diplomacy, competition, and peace are at the extreme of this spectrum and as conflicts intensify and propaganda and rhetoric increase, those who wish to make war create conditions that tend to exacerbate the problems. Economic conditions also drive populations to force their leaders to identify causes and assess blame. And in some cases, discrimination or other conditions drive harsh realities toward intensified conflict.

3.6 War and all out use of force

As conflict levels increase, the intensity and use of force also increases. From diplomacy the typical path goes to economic and political leverage, then to posturing, supporting the enemy of your enemy, shows of force, and finally, to war.

Economic and political leverage include things like sanctions enforced on a global basis and the use of influence to inhibit or punish actions of the enemy. This happens in all sorts of ways, from the senders of spam emails who are listed as spammers and blocked from sending further emails to protesters sitting outside the seats of power. People tend to use their influence, tell friends and family, threaten loss of business, and eventually, they may carry out those threats.

> Violent acts send a clear message, often leading to more violence.

Posturing is like the chest beating seen in apes of various sorts. It involves showing that you are willing to take actions if necessary. Saber rattling prior to battle used noise to literally rattle the enemy. The goal is to get the opponent to back down. Of course sometimes this is simply done to appease an audience at home, showing that you tried without really trying all that hard, to threaten rather than use force. And of course the ancient Arabic saying "*The enemy of my enemy is my friend*", except of course that many times conflict involves more than two parties. More on that later.

At the extreme end of the spectrum there is war, a series of forceful exchanges in which violent acts are used to compel them to do what we want. Of course there are many types of force. For example, when we take actions to sever all communications and stop the flow of food and electrical power, this may produce the same effect as explosive devices, perhaps even more death and suffering, and more internal discontent. If we do these with explosives it has a different psychological impact than doing it by Internet attacks, hence terrorists prefer blowing things up.

3.7 The defensive picture

The defensive picture is depicted here in table format. This table is certainly not complete, but it gives a sense of the sorts of options available in different situations. The vertical is the dimension of low to high frequency, covert to overt, and strategic to tactical with "maximum" indicating high frequency, overt, tactical and low indicating low frequency, covert, and strategic. The horizontal is the intensity dimension with low to maximum elements indicated by the bottom row.

Freq\Inten	Low	Medium	High	Maximum
Maximum	War games	Increase self-test	Enhance and reconfigure infrastructure	Respond and rebuild
High	Increase posture	More response resources	Limit non-critical uses	Anti psy-ops
Medium	Assess posture Expose enemy deception	Reduce detection thresholds Control technology use		Limit noise sources
Low	Develop knowledge			Increase operations security
What	Diplomacy	Economic and political leverage	Show of force Posture Support enemy	War

The frequency vs. intensity space for defense

Starting at the bottom left, and moving toward the top right, the indicated actions take place as intensity of conflict increases and get turned back off as intensity decreases, with the things lower and to the left taking place at higher frequency and intensity levels as well, when feasible and appropriate. This escalation and deescalation process is how conflict intensifies and eases off in the times building up to and backing down from war.

The defensive elements indicated include:

- **Develop knowledge:** This activity is the eternal intelligence and educational effort required in order to prosper in conflict. More and more accurate information that can be used to prosper is always desirable.
- **Expose enemy deception:** Exposing the deceptions of your enemy helps to mitigate the perception management efforts they undertake. Public disclosure is not always desirable because it provides information to the enemy on your capabilities, however occasional exposure can be used for gaining political support at home.
- **Assess posture:** Knowing the status of your own protection program is fundamental to being able to adapt to conflict.
- **Increase posture:** When conflict starts to increase in intensity, it is important to increase the level of vigilance spent on defenses. But the cost of increased protection posture can be substantial, so it is normally optimized for the situation at hand.
- **War games:** Practice makes perfect. War gaming is about testing scenarios, practicing how to act under real-time attacks, and anticipating a range of scenarios so that you don't end up with failures of imagination.
- **Control technology use:** Controlling the use of your own technology and limiting the technologies of the enemy are approaches that allow technological advantage to be gained and retained. By limiting your own use, you can preserve more fragile elements. By limiting enemy use you can reduce their ability to attack and defend.
- **Reduce detection thresholds:** Detection typically gets triggered based on reaching a threshold of urgency, and import. As intensity increases, the level at which you decide to respond typically goes down, which is to say, you become more sensitive to events and pay more attention to anything that you can see happening.

- **Increase response resources:** The decrease in detection thresholds also implies an increase in resources needed to deal with the event sequences detected.
- **Increase self-test efforts:** As intensity increases, the need to test yourself and make certain you are prepared also increases. War games become more intensified and people are willing to explore issues that they tend to ignore in peaceful situations. Presumptions of malice instead of benevolence become assumed.
- **Limit non-critical uses:** Non critical uses are often eliminated to assure that adequate resources are available for critical uses. Strangely this also tends to lead to less accountability because accounting for actions takes time and effort that does not directly impact results. For many, war is an opportunity to steal.
- **Enhance and reconfigure infrastructures:** Because normal operations on infrastructures tend to be based on normal load levels and increased intensity increases those load levels, infrastructure elements are often added even as normal loads are shed. Reconfiguration is sometimes used to defeat previous enemy intelligence efforts so that the information they had before the war doesn't work during the war. Of course it may also tend to confuse your side if not carefully done.
- **Increase operations security:** Operations security is necessarily increased as the number and consequences of operations increase. "*Loose lips sink ships*" may become the watch phrase of the day.
- **Limit noise sources:** Many of the everyday things that occur are more or less noise compared to the high implications of losing a war. So these sources may be cut off for a time to allow better focus of attention.
- **Counter psychological operations:** This means finding ways to detect and defeat psychological attacks against you, hopefully without inhibiting your ability to further detect them by revealing sources and methods.
- **Respond and rebuild:** Keep fixing whatever they break.

3.8 The offensive picture

The offensive picture operates in much the same way as the defensive picture. The dimensions and parameters are the same, but the actions are on the attack side rather then the defense side.

Freq\Intens	Low	Medium	High	Maximum
Maximum	Probe competitors	Weaken enemies	Weaken Infrastructure Deception	Destroy infrastructure
High	Test for weakness	Increase noise Destabilize beliefs	Exercise planted weaknesses	Exploit Psyops
Medium	Plant weakness	Exploit known weaknesses	Exploit more known weaknesses	Exploit all known weaknesses
Low	Control R&D	Limit technology use		
What	Diplomacy	Economic and political leverage	Show of force Posture Support enemy	War

The frequency vs. intensity space for offense

Again, we have the escalation and deescalation process supported by a range of approaches, again everything lower and to the left continues to a greater or lesser extent during more intense conflicts, and again this represents only a limited subset of the actions that take place in the information arena. The offensive elements indicated include:

- **Control research and development:** Just as the defensive side seeks to use research and development to advantage, so does the offensive side. For example, by limiting research in certain areas of biology, economic and cultural advantages can be gained. Whether we are talking about the Chinese limiting the US efforts in biological research or religious groups limiting stem cell research, they are both examples of information warfare.

- **Plant weaknesses:** Intentionally planting weaknesses has been a long tradition in information protection as well as many other arenas. The Trojan horse concept is applied in computer hardware and software, in design specifications, and in any number of other ways to intentionally induce subtle weaknesses that I can exploit on demand to attack you. These tend to be strategic because of the long time frames involved in planning, implementing, and planting undetected Trojan horses.
- **Test for weaknesses:** Planted and natural weaknesses have to be identified, tracked, and verified prior to use except in cases where they are simply statistical parts of larger noise-level attack strategies. If you depend on it for advantage in a situation, it had better work, or the advantage is lost.
- **Probe competitors:** Since you never know who you will be going to war with next, all likely competitors have to be probed continuously in order to be prepared for war. The probes produce intelligence information that is used in making tactical decisions. The information from probes is aggregated, fused, analyzed, and assessed in order to produce intelligence products for higher and higher echelons, ultimately producing national-level intelligence reports for leaders of nation states and other intelligence products for different customers. In the Internet era, probing competitors has reached incredible proportions, producing detailed information on individuals and their capabilities, intents, location, lifestyle, family, food, friends, things, people, activities, everything they have every written or said, and so forth.
- **Limit technology use:** The offensive goal is to limit the enemy's use of technology so that they lose the advantages that technology brings. The history of war is literally littered with the bodies of those with inferior technology. Technology and training have reduced the loss of life to US troops in battle to 1% of the enemy losses from almost 100% only 20 years ago.

- **Exploit known weaknesses:** At some point exploits are undertaken to take direct advantage to weaknesses found in probes or planted by strategic efforts. Widely known weaknesses are typically exploited or at least exercised more than others at lower levels of intensity to prevent loss of the more expensive and covert resources through their exposure to scrutiny.
- **Destabilize beliefs:** While defenses tend to seek to solidify support and beliefs associated with the war effort, on the offensive side, the goal is to destabilize the belief in the leadership and/or the cause of the opponent. Whether it is political propaganda or the Voice of America during the Cold War, propaganda, seeding dissent and discontent, creating artificial enemies, and reducing trust in leadership are fundamental aspect of this element of the offense.
- **Increase noise:** The level of general events keeps most defenders busy during low intensity conflict, but part of the object of offense is to destabilize the defender's capability. This is done by, among other things, increasing the level of noise and spurious things that the defender has to do. This also helps to conceal the surreptitious acts associated with long-term planted capabilities so that they are harder to find among the din of other similar things underway.
- **Weaken enemies:** Generally, anything that can be used to weaken enemies, from reducing food supply to increasing prices of goods, to poisoning water supplies might be tried. Of course some of these are considered acts of war, some are against globally recognized acceptable conduct and risk getting the whole world lined up against you, and some are considered acceptable without or before a formal declaration of war.
- **Exploit more known weaknesses:** As the intensity level increases, more enemy weaknesses are exploited. During open hostilities most widely and some less widely known weaknesses are exploited. Some will be kept in reserve.

- **Exercise planted weaknesses:** Planted capabilities such as Trojan horses and similar mechanisms are exploited at increasing intensity as the conflict escalates. These are typically classified military secrets and the capabilities may be lost if used, but at some level of intensity, the risk of exposure is less important then the utility they bring, or there would be no rational reason for planting them in the first place.
- **Apply deceptions:** Deceptions of all sorts are applied, including outright lies, misdirection, tactical deceptions related to maneuver, feints, camouflage, and everything else available. The typical goal is to gain a short term advantage so as to win a battle or gain some other advantage for the short term.
- **Weaken enemy infrastructure:** Enemy infrastructure attacks are typically designed for optimal exploitation, which includes not only denial of services, but intelligence gathering, propaganda, financial attacks, supply and logistics attacks, and all manner of other attacks that get at the will and ability of the enemy population and leadership to fight and support the war effort.
- **Exploit all known weaknesses:** At some point any and all weaknesses known are exploited as well as can be done and in coordination so that exploitation does not destroy other capabilities needed for the war effort. For example, disrupting the computers being used to disrupt your economy means you are attacking your own capabilities instead of your enemy's.
- **Exploit psychological operations:** Information resources and enemy mindset are exploited to destroy the will of the enemy to fight. Whether it is false letters from home telling the fighters of the suffering and loss of their sons, or pamphlets explaining how to surrender, it is in use at this level of intensity.
- **Destroy and disable infrastructure:** Enemy infrastructure is destroyed or disabled, depending on the military and post-military situation desired by the attacker.

3.9 Escalation and deescalation

The movement of conflict up and down the intensity scale is almost always accomplished by intentional acts on one side, followed by intentional responses by the other, and so forth. While escalation in a bar fight comes from instantaneous impulses of the actors over very short time frames, larger organizations escalate more slowly and when more forces are involved, it takes time and effort to make things happen.

Escalation usually involves a series of tit-for-tat executions on one side then the other, but in some rare cases, notably what is globally referred to as aggressive war, one side just comes out and attacks the other without an escalating series of events. This sort of action is widely frowned upon and it takes a much bigger propaganda campaign to cause it to happen without enormous resistance. After all, who wants to be a pariah to the world? Examples of thinly veiled aggressive war of this sort are Germany and Japan in World War 2, and the US in the invasion of Iraq. While each aggressor cited causes and made claims of injury or potential threats of injury, each acted without a series of increasingly intense exchanges.

Deescalation is far harder for waring parties to undertake and, as a result, most wars start rather more quickly than they end. This is good for the egos of the parties who can claim some sort of victory, or at least save face in some way. But it is bad for the people of those countries who have to suffer loss of life, economic hardships, destruction of home and property, and all of the other slings and arrows of outrageous fortune so that their leaders can feel good about themselves.

Not all processes go in one direction then the other. Many go up and down the intensity scale. Pakistan and India have been in conflict for many years with escalations and deescalations all along the way. This ebb and flow of intensity sometimes reaches equilibrium for a while, but because of the cost of war, it tends to naturally deescalate rather than remain stable at high intensity.

3.10 These aren't just theories

I know I haven't given examples of each and every one of these things, but this does not mean that they are in any way theoretical. In fact, each and every one of these things is done in conflicts all of the time. I will try to be more documentary and give examples of these throughout the rest of the book, so for now just buy into my premise, or go research one of them to prove me wrong.

3.11 Coming to peace

In the end, the goal of every rational human on Earth should be to come to peace with everyone around us, to raise all boats to a level where the life and death struggles that lead the desperate to war disappear, the 40,000 people a day that die of starvation are fed, the religious zealots that call for death to everyone who doesn't agree with them are marginalized and ignored, and humanity moves into the age of enlightenment and environmentally sound building instead of the age of destruction and distrust.

Peace is not a stable state.

But this is not exactly likely to happen soon. So many of us might be willing to settle for a situation in which we aren't killing each other over money and oil, in which our children are not going off to kill others and die themselves unless and until it is absolutely necessary. Or how about a world in which the techniques of information warfare lead to dismantling of violence through management of perceptions around the globe? This is also unlikely, and there is a very good reason. Peace is not a stable state.

If you think that humanity can reach a state in which we are all at peace with each other, as much as I wish it were not so, you are wrong. The reality of the human condition and the condition of all living creatures on the Earth is that life is competitive by nature. Disease emerges and kills millions. People get injured and other people resent having to work to take care of them. A greedy person in a world without greed prospers and others get jealous. Peace, justice, and equality can never be perfectly attained. There will always be conflict because people are not perfect.

3.12 Individualized conflict and resolution

As technology increasingly moves toward the ability to manufacture or deliver one of almost anything to almost anywhere in almost no time, the notions of conflict increasingly become individualized and in some cases reach the individual level.

This book is an example of such a technology, and my series of security awareness books exemplify its use in commercial venues. This book is produced in the UK and US in quantities as low as 1 for about the same unit price as I was able to manufacture 5,000 books only ten years ago. That means that when someone requested a copy of my last book in France, I had it manufactured in and shipped from England the next day and had it to them before I could have shipped it to him in Europe if I already had a copy made.

My awareness books are made custom for each enterprise customer. They order up the elements of an awareness program that they want and get a draft copy within a few hours. They approve the final or make changes, and when they are good with it, I manufacture the number they want to purchase customized to their enterprise for a cost below what they could buy the non-customized book for at wholesale. Consider the extensions of this to other applications.

> Democracy depends on a well informed public. If I can tailor a message to each individual, I can focus on the things they like about me and the things they hate about my opponent.

Politicians are on the leading edge of individualized marketing. The ability to use databases for targeting specific individuals with just the right information has become the hallmark of the modern political campaign. If you like hunting for deer and chocolate ice cream, I can get you the political message that will enamor you to me and your neighbor a completely different one for almost the same cost as sending out a mass market mailer with identical content. This may seem frivolous, but if you think through it, it means the end of democracy as we know it.

The ability to get individualized messages to individuals at a wholesale level means that instead of being two-faced, I can be million-faced. And so can you. But at increasing levels of intensity, this individualization of everything has enormous implications for conflicts. It means, at the violent extreme, the capability of killing you and your family from afar and with no reasonable chance that you will have any indicators in advance. To defeat this, it becomes necessary to use surveillance at unprecedented levels. And who runs the surveillance system? Since governments have a long history of abuse of surveillance, this leads to the return of the junta with leaders and aides and all of the other elements of the classic military dictatorship, but with one big difference. They will have the intelligence capability needed to enforce their rule with iron fists.

I have personally been involved in Internet-based conflicts where individuals and small groups go after other individuals and small groups with intensity levels ranging up to nearly the same levels reached in military information operations. While the scale may be a bit different, it is not dramatically so. I have been in battles involving scores of actors from all over the world. Some launched direct or indirect distributed coordinated attacks against my information infrastructure, some leveraged hundreds of US military systems, some involved thousands of systems, and I have helped others in similar battles. Some of these battles were part of terrorist actions, some involved nation states actively attacking each other when there was simultaneous violence between them, and some involved possible precursors to nuclear conflicts.

I have also helped other investigators in tracking down individuals who ended up being real terrorists planning attacks within the US, been joined in battles on behalf of corporations involving foreign intelligence agencies, and I have been in personal one-on-one conflicts with individuals who were personally mad at me for one reason or another, rational or not. Not all personal battles are military or violent in nature, and economic battles, personality-based battles, and any number of other less violent acts of aggression persist in the human condition.

4 Information war and nation states

Nation states participate in information warfare and have for a long time. In addition to the use of propaganda and the other tools of the state, or before that, tribes and lesser organizations, direct conflict between nation states in the information arena has become far more prevalent and dangerous over recent years because, at least in part, the increased dependency of technologically advanced countries on information and information technology for day-to-day life.

> Nation states have brought death and destruction to the world by their use of industrial age techniques in conflict and warfare. And as nation states enter into the information warfare arena, they will increasingly bring information age techniques to warfare causing a different sort of global chaos and mass disruption.

Conflicts between nation states have historically been the most violent, far reaching, and dangerous of the conflicts faced by humankind. At the high intensity extreme, we have the world wars that cause global chaos, kill tens of millions of people, harm billions of people, and devastate economies, destroy historical remnants turning them to rubble, wipe out large bodies of human knowledge, displace many millions of people and disrupt their lives and families, create conditions that foster hatred across racial, religious, and other lines, and generally wreck havoc on the world.

Nation states also play in the information warfare arena, and of course their conflicts are far longer term, more consequential on the large scale, and harmful than the conflicts of smaller bodies. At least for now, they are somewhat restrained, but don't imagine it will last for long. Increasingly, information in warfare is being used and information wars are underway already, even if some of the participants are unaware of their involvement in it.

4.1 The US

The United States has been a global leader in information warfare, or at least they think they have been, for some time. Unfortunately for the US, they are apparently unaware that they are not the world leader in information warfare and that they are and have long been under attack and are in fact losing the information war. The reason they can't seem to get it is that they are so egotistical that they refuse to believe that they could be second to anyone.

The US makes wild claims about being the only superpower, and yet they are in one major conflict in a few countries in the Middle East and they cannot project power effectively to anywhere else in the world. They try to use their military might to project power but fail to realize just how vulnerable they are to strategic warfare or even that such warfare exists. It does, the US doesn't get it, and they are losing.

> After securing what seemed to be an absolute victory in the information arena, the US managed to make what might be the single greatest strategic blunder of the age when they literally threw away their global leadership position for a generation to come.

From an information warfare standpoint, the US has enormous offensive capabilities when it comes to leveraging IT for faster tempo, accuracy, and efficiency in the physical battle space. But as the leaders of the world in information technology since the 1950s, the US somehow managed to decide to stop funding much of the research in this area in universities in the late 1980s as part of the reduction in educational research funding by the Federal government. The result was a reduction in the number of graduates, amount of research, and innovation in information technology at the key moment in history when it was most needed – the run-up to the information age, the Internet revolution, and the massive globalization associated with communication and information replacing transportation for much of human endeavor.

The result was devastating. Much of the information technology base associated with hardware moved offshore, research into information protection was cut to nearly zero, so the US created technology that was indefensible, the education of the next generation of IT experts never happened, and at the same time, the rest of the world came to the US to get educated and create their expertise in this area. They generated the graduate level expertise they needed as seed corn to create their information industries and their educational systems, and they now outstrip the US in number and quality of graduates, production of software and hardware, advances in technology, robotics that helps to automate manufacturing processes, and so forth. In other words, the US built the capacity to win the global information war for a generation to come and then literally threw it away to save a hundredth of a percent of the Federal budget that went to pork barrel politics.

The lead that the US had built up lasted a short while, but as it collapsed along with the Internet bubble in the late 1990s, the US found itself in an ever worsening position, outsourcing and offshoring the IT capacity of the country, reducing the number of graduates in IT, especially at the graduate level, educating an increasing population of people from other countries that were to return home with the knowledge they acquired instead of retaining it in the US, and facing an aging population of professors not being adequately replaced. So as the world rose, the US fell, thinking all along that they were unbeatable and that economic forces would somehow balance all through free trade, when the resulting balance was to reduce the value of the US and raise all other boats around the world.

> The loss of seed corn is the worst part of the US lack of strategy since it makes rebuilding the iwar capacity all the harder.

While the US tossed out its strategic lead in IT, it didn't abandon its tactical interest or efforts. The ongoing research and development efforts undertaken by the US in the direct application of information warfare had substantial increases throughout the 1990s leading to the further development of special weapons and tactics that enhanced and improved upon its capabilities for offensive information operations, perception management and propaganda, network centric warfare, technical intelligence gathering, and many other aspects of information warfare. These continued to provide progress at a high rate, but the lead is decreasing today.

The turn of the tide is underway and the US has largely burned out its capacity for nation building with its efforts in Iraq and Afghanistan. While the US has enormous technical information surveillance assets, its human intelligence capabilities were degraded year by year, turning into an inability to detect strategic attacks like those on September 11, 2001 because its technical capacity could not deal with sub-state actors. And all of this drew the attention of the US away from other serious threats like China and Russia and the indirect effects of loss of technical and manufacturing capacity on the underlying infrastructures that support information warfare and national capacity.

> The US has ceded its lead and outsourced its very survival to its enemies.

The most devastating aspects of these losses lie in the elimination of the capacity to build its own information technology and infrastructures. The shift from US-based manufacturing of information technology and the components that comprise its own infrastructures has created the capacity for enemies to plant Trojan horse hardware and software embedded throughout the entirety of US infrastructure. From water supplies to electrical power to computers and communications systems, to databases, to every other facet of the underlying technologies that support transportation, communications, finance, energy, emergency response, medicine, and even the basic sciences, the US has ceded its global leadership, not only to the rest of the world, but to its direct nation state competitors. While the US has built up its offense, its defense is essentially void, its ability to move quickly in research and development gone, and the fall nearer than it imagines. But a fall to whom?

4.2 China

The most likely candidate for defeating the US in information warfare is China. And, not surprisingly, China declared its intent to pass the US in economic and information arenas. In the late 1990s, China indicated that by 2005, they would be positioned to win information conflict against the US. I wrote a short thought piece late in 2004 asking how close they were to winning and, not surprisingly, they were doing rather well. I have extracted parts of it here for your interest.

...What would you consider a victory in the information war if you were China and a loss if you were the US? I'll make a little list of some things that will, of course, be missed...

- **China Win:** China's economy outstrips the US to the point where they have more global influence than the US. They are just about the same size as the US today and growing while the US economy is slipping despite what the US officials say.
- **China Win:** Better and more higher tech jobs in China to the point where they start to dominate global IT from design to manufacturing of hardware and software and eventually systems of all sorts. Let's see. In 2004, Cisco announced they are going to move to China in a big way, Microsoft licensed or moved much of it's IT capabilities to allow China to work on it, Linux and other open source is widely adopted in China as the gold standard, chip designers have to go to Chinese cryptography, they maintain content control over what enters their market, and their market dominates.
- **China Win:** Costs go up fast in the US to the point where economic growth is stunted, and supplies of oil and other critical support systems fall increasingly under Chinese control. China is next to middle eastern oil and can create pipelines to move the resources into China for less than the US, meanwhile, US supplies are disrupted, costs skyrocket, and China thrives while the US survives.

- **China Win:** China builds up military capabilities in a wide variety of areas without US attention or ability to influence it significantly. Clearly the situation today has the US focused so far away from China that they are building up a naval capability, an ability to take over Taiwan - if and when they choose but no reason to really push it, and they build military information warfare capabilities and test them in the US with plausible deniability.
- **China Win:** The US buys lots of goods from China, boosting their economy while the US runs a trade deficit and the Chinese buy up US assets at low rates, ultimately coming to control a significant portion of US assets and influence greatly US companies and business directions. Look at the figures and you will see all of this in place already with more on the way.
- **China Win:** The US is completely distracted by other events so it doesn't notice or act to stop the Chinese strategy. The US military is depleted to the point where it really cannot act effectively on the same global basis it used to, and with decreasing morale in the US military and nation.
- **US Win:** ... Uh oh...

So if you look at it, it looks to me like China is not only winning, but they have very nearly won. The die is cast so to speak. Within a few short years, the US will potentially be starved for oil if the Chinese choose to make it happen. The US information capabilities are increasingly under control or influence by China and they have access to the ability to plant Trojan horses and find vulnerabilities at will in most US systems. They have ownership levels that will soon reach the point where they can control large portions of US business decision-making, and exert political influence through US nationals who run those companies and have great influence.

The Chinese have more economic clout around the world. They are increasingly gaining more political clout throughout the world. The US is destroying its political and economic clout at the same time.

World War 3: We're losing it...

The high tech jobs are going to China so they have the strategic long-term capability that the US is throwing away. They have a larger military that will eventually exceed US capabilities as it goes unchecked year after year. And the US doesn't even see it coming.

That's what I wrote a year or so ago, and of course things have been making progress since then. For example, China bought out IBM's laptop PC line so they are now building more of the computers used in the US than anyone else. China now builds more than 70% of all computer circuit boards, has been growing at more than 29% per year in its microelectronics business, and is growing faster than the rest of the world in almost every area you can identify in terms of information technology. Leading global microelectronics manufacturers with large Chinese manufacturing and technology capabilities include Celetica, Flextronics, Jabil, Sanmina-SCI, and Solectron. In fact most of the world's largest microelectronics firms are rushing to China for the low labor costs and access to the enormous emerging market. For more details, read http://www.buyusainfo.net/docs/x_3591025.pdf.

If any country is poised to win the strategic information war, it is China. Or that's what you might think from looking at pure efforts to produce the basic components. But China does a lot more than that. Their corporate and military espionage are top flight. The US has recently arrested several of their agents for theft of US military secrets after they stole detailed plans for classified US information and electronic warfare technologies. They also have demonstrated infiltration of operatives into critical infrastructure control systems including but not limited to power and telecommunications, the two most critical elements necessary for operation of information and information technology capabilities, the supporting infrastructures for finance and business operations, and of course absolutely critical elements of US information warfare.

> China has a long history of corporate and military espionage against the US and is making incredible progress in taking over the basic building blocks of national infrastructure.

While some may attribute these items to normal economic competition and the emergence of China from a closed society, this is naive in the extreme. According to Ravi Prasad's, work in the early 2000 time frame, In mid-1999, China established a special task force on information warfare composed of senior politicians, military officers and academics, headed by Xie Guang, Vice-Minister of the Commission of Science, Technology and Industry for National Defense. This task force prepared detailed plans to cripple the civilian information infrastructures of Taiwan, the US, India, Japan, and South Korea. Two members, Qi Jianguo and Dai Qingmin, have formulated a comprehensive scheme:

> China has a well defined strategy and doctrine to target the US, India, Japan, and South Korea.

- China will not attack military or political targets in these countries but will target their financial, banking, electrical supply, water, sewage and telecommunications networks.
- Chinese companies will establish business links with private companies in these countries. After carrying on legitimate business for some time, they will insert malicious computer codes and viruses over commercial e-mail services.
- The viruses and malicious codes will be sent through computers in universities in third countries so that they can not be traced back to China.
- The attacks will be launched when the political leadership of the target countries is preoccupied with election campaigns.

The People's Liberation Army (PLA) has conducted several field exercises in this area. An *Informaticised Peoples' Warfare Network Simulation Exercise* was conducted in Echeng district of Hubei province. Five hundred soldiers simulated cyber attacks on the telecommunications, electricity, finance, and television sectors of Taiwan, India, Japan and South Korea.

Ten functions were rehearsed in another exercise at Xian in Jinan Military Region: planting information mines, conducting information reconnaissance, changing network data, releasing information bombs, dumping garbage, disseminating propaganda, applying deception, releasing clone information, organizing information defenses, and establishing network spy stations.

In Datong, 40 PLA specialists are preparing methods of seizing control of networks of commercial Internet service providers in Taiwan, India, Japan, and South Korea. They held demonstrations for the Beijing Region Military Command, Central Military Commission and General Staff Directorate.

China is very serious about full spectrum non-violent warfare and, based on their identified plans and efforts, current events support a lot of progress in meeting their strategic goals.

In October of 2000, Chief of General Staff Fu Quanyou presided over an exercise in Lanzhou and Shenyang Military Regions which simulated electronic confrontation with countries south and west of the Gobi Desert. This focused on electronic reconnaissance, counter-reconnaissance, electronic interference and counter-interference. It tested the battle readiness of PLA's command automation systems, command operations, situation maps, audio and graphics processes and controls, and data encryption systems. Smaller exercises were carried out in July in the Chengdu Military Region and in August in the Guangzhou Military Region.

The PLA has also enlisted support from universities.

- It established the *Communications Command Academy* in Wuhan in collaboration with Hubei's engineering universities. The *Navy Engineering College* collaborated on secret projects on information warfare with them.
- The PLA established the *Information Engineering University*, headed by Major General Zhou Rongting, in Zhengzhou by taking over and combining the civilian

Institute of Information Engineering, *Electronic Technology College* and *Survey and Mapping College*. They specialize in remote image information engineering, satellite-navigation and positioning engineering, and map data banks of the regions from India to Indo-China.

- The PLA established the *Science and Engineering University*, headed by Major General Si Laiyi, by combining the civilian *Institute of Communications Engineering*, the *Institute of the Engineering Corps*, the Air Force's *Meteorology Institute* and the *Research Institute of General Staff Headquarters*. They attracted more than 400 civilian professors from universities all over China, established of a new *Institute of Computer and Command Automation,* and persuaded 60 experts of Chinese origin settled in the West to return to work there.
- A fourth PLA institute is the *National Defense Science and Technology University* in Changsha, under direct supervision of the Central Military Commission, where the Yin He series of supercomputers has been developed.

Make no mistake about it. China has a well defined strategy, and doctrine, and has taken specific actions to achieve success in information warfare against specific targets in specific nations. Its recent moves against US information technology companies cannot be seen as happenstance. It is leveraging its low labor costs, not just to bring its people out of poverty, but to bring US, Indian, and other select nation states under more direct control of its influence. Its moves to gain contracts for telecommunications that dominate international links between countries is also a strong indicator of the success of their efforts. They are also making political moves in South America where they are starting to create alliances with countries that are increasingly distancing themselves from the US. They are using their vast resources to support corporate moves that place them in positions of information dominance and they are doing this all over the World.

4.3 India

India is and has long been a US ally, is a rare democracy in their part of the world, and had long been one of the greatest supporters of the US in that part of the world. Of course the same cannot be said about the US as a supporter of India. The US has not supported India, preferring to remain neutral in the disputes with Pakistan, and was greatly offended when each nation became a nuclear power.

> India's investment prior to Y2K clearly showed the value of offshoring IT to India.

But in information war, India has a big advantage today because of its long-term investment in information technology. India started seriously investing in information technology and electrical engineering as well as other aspects of science and engineering in the middle of the 20th century, as did the US. But when the US reduced its investments in long-term research in the 1980s, India expanded its efforts and expanded the number of advanced degrees generated and the level of scholarship while producing increasing numbers of professors, masters student graduates, and bachelors degree graduates. Taking advantage of the US scholastic institutions, they educated tens of thousands of graduate level engineers and programmers and started an enormous industry.

The next strategic step was the offshoring offerings from India that took advantage of English as a major language in India for services like telephone support, technical support, and other related functions. Offering services at lower prices and taking advantage of its increasing investments in information infrastructure, India began its ascent in the information arena with assistance in change-over efforts for the year 2000. By providing many thousands of programmers at reduced cost, they took large numbers of programming jobs in the US, created the infrastructure and business relationships required for sustained growth, and simultaneously planted software Trojan horses in US critical infrastructures.

As India built business relationships with its US customers, the reduction in costs combined with the ability to deliver on information technology projects led to increased movement toward offshoring US high-technology and information-related jobs to India wherever possible. The combination of a highly stable democratic government, enormous IT expertise, English speaking, outstanding information infrastructure, and growth of a customer base of a billion potential customers led many companies to move whatever functions they could to India, as they are still moving today.

In India, some things are very similar to the US and some things are very different. In particular, there is a very serious cultural difference in the creation of certifications, the liability laws, the legal system in general, the caste system, and the way business is done. For average quality and low risk, India does a great job of supporting US information technology and content delivery requirements. But as the surety requirements increase, the disadvantages of inadequate personnel background checks, the need for trusted individuals to develop national security systems, and the requirement of strong testing and assurance programs tends to still, but barely, favor US top-end developers. But this will not last long as India seeks to improve quality and increase its utility and as the US continues to squander whatever capabilities it has in this arena.

If it were only information technology, the US might claim some advantage over India, but they have also recently started to demonstrate leadership in other areas of technology, science, and engineering. For example, they have recently started to offer lower cost top quality health care that includes far nicer conditions than US hospitals and combines them with luxurious recovery facilities. How long will it be before they surpass the US in biotechnology? Not long. And the same goes for engineering in other areas.

> India's high-tech is now surpassing the US and in the information arena, they have many advantages that give them advantages in the iwar arena.

4.4 The Former Soviet Union and Russia

The former Soviet Union was largely ruled from Russia, and even as they collapsed, among the few bright spots for Russia was its intelligence organization and its skills in information technology, and that included very strong capabilities for attacking information technology and systems as well as mathematical skills that are world class. And when the Soviet Union broke up, those capabilities did not disappear, rather they reformed into what might be a far more dangerous set of capabilities.

> Russia remains a formidable information warfare threat, and perhaps even a more diverse one than when they were tightly controlled under the former Soviet regimes.

The unified threat of information warfare has not completely disappeared and much of it stemmed from Russians who went forth and started their own independent operations. Some went the criminal route while others went the government route and still others went into the security business. Those that went into the security business make modest livings protecting other people and their businesses, do research and development of defenses, and so forth. But for those who specialize in offense, defense is not a game that is easily undertaken. They have spent their lives understanding how to find niche weaknesses and develop them into leveraged advantage for targeted attacks. To become broad spectrum experts in a wide array of specialties takes many years and you have to put food on the table along the way.

It is far easier for an attacker to leverage what they know to break into systems than to try to help others defend them. From joining forces with the Russian mafia, to striking out on their own, to forming intelligence companies to do competitive intelligence, to going into the extortion and bank robbery business, these ex-information warriors have formed the Russian irregulars that eat away at system after system without discernible controls. But this is not to say that the government controlled elements are gone.

Indeed they are not. Russian attackers have broken into systems all over the world for intelligence operations and on the rare occasions when traces are found of leaking information from classified systems through Russia, they plausibly deny any knowledge claiming that someone broke into their systems while providing absolutely no assistance and a great deal of cover along the way.

The most famous cases are the ones published in the papers every once in a while where the US has been looking for Russian attackers that got into DoD computers and have been systematically leaking information for years undetected. They are accidentally detected by someone in the DoD somewhere while looking into an anomaly on their own, and turn into major affairs. They are usually Trojan horses that exploit covert channels at low bandwidth to leak highly confidential or even classified information. The overall intelligence operations and capacity for large-scale sabotage still continue unabated.

Russian owned companies in the competitive intelligence business are quite active and have demonstrated skills and willingness to break into companies to gain information. This is illegal, but if properly shielded through national boundaries the risk is minimal to non-existent. Just make sure to pay off the proper folks.

> Russian electronic warfare capabilities are very strong. They have demonstrated ability to disrupt battlefield information operations.

Want an electromagnetic weapon? The Russians were outstanding in this area and for the right amount of money, you can almost certainly purchase a disruptive capability that can be aimed at a building from outside the walls to disrupt computers within. Of course this can be defeated, but only by proper physical security design, which most enterprises lack. Electromagnetic pulse weapons for the battlefield are also a Russian specialty and could be used to defeat the high technology high tempo warfighting used by joint and allied forces in so many recent battles.

4.5 France

France has long been in the information technology arena and has had a strong program since its revitalization after World War 2. Over the years, France has made some key decisions that have clearly been related to its capacity to win in the information arena.

For example, France restricts the use of cryptography in communications and requires that the French government have the capacity to decrypt all communications. This creates problems for international banking which relies on encryption for secure transfer of electronic funds. It also allows France to do unfettered surveillance on businesses and individuals communicating within or through France. And France put in a very large communications infrastructure to support communications between adjoining countries, complete with restrictions on the use of cryptography.

France is not as formidable in iwar as they wish they were.

This is seen by many as part of the French approach to financial survival and business dominance. Industrial espionage, theft of data from computers of executives on airlines, sale of technologies to other nations, and related issues have come up again and again with France. For example:

- On French airlines, they take computers from passengers and duplicate hard drive contents for industrial espionage.
- French companies have long been involved in the sale of military technology to nations that have been under UN sanctions.

But despite some well publicized incidents, the information warfare capabilities of France have not brought it to a leadership position in the world. While it sells nuclear power technology and develops its own information technology, its warfighting capabilities and exploitation of information seem highly limited relative to the big players in the international information warfare community.

4.6 Canada

Canada is used here as an example of a nation that is not particularly war-like and yet has a strong program in information warfare. The Canadian intelligence organization and diplomatic corps, the Communications Security Establishment (CSE), and its joint warfighting capabilities with the US, all demonstrate a level of sophistication and understanding that are often lacking in larger military organizations.

Canada has shown a strong interest in information warfare for many years. They have their CSE that roughly corresponds to at least the defensive information operations capabilities of the US National Security Agency. The CSE promulgates and reviews hardware and software for meeting standards that are rigorous and meaningful for defending national information assets.

> Canda's CSE is roughly equivalent to the US NSA in terms of its role in defensive information operations. Its offensive capabilities are closely held secrets.

With lesser dependence than many countries on information technology for economic well being and a more self-sufficient citizenry, Canada has relatively little at risk from information attack. While it is certainly a modern nation, it has the North pole on one side, the US on the other side, and oceans on the remaining sides, all of which act as very effective buffer zones against direct military conflict. This means that Canada only really has to deal with the information warfare issues when it comes to political subversion, economic attack, propaganda, and projection of force. Canada's projection of force is relatively small and more or less dedicated to maintaining peaceful and diplomatic relationships while remaining friends and allies with as many countries as feasible. It integrates effectively with US and joint operations as needed but does not have a massive military infrastructure like the US has.

Canada has issues with sub state actors such as very small groups of terrorists that house themselves in some Canadian provinces and Chinese information warriors that base themselves in Canada at times. But these are not aimed at Canada. They primarily act as a staging area for attacks on the US.

As a result of these various factors, Canada needs and has a stronger defensive than offensive capability in information warfare. Propaganda is not as deeply applied in Canada as in the US or other emerging nations that apply these tactics against their own citizens, but there are substantial internal threats associated with French and English language differences and the desire of one province to cede from the rest of the nation. These sorts of information issues create internal problems and act as points of friction for the country and its political system.

> Canada is particularly good at remaining friendly with everyone even though they are actively exploiting them all in the information arena.

Canada has information operations on a global scale. They have diplomatic missions all over the globe and often end up secretly housing dissidents or intelligence operatives of other countries in Canadian embassies, acting as surrogates for communications, broker deals using their contacts, help to extract information that could not otherwise be taken out, and so forth. They maintain a global reputation and appearance of a peaceful fair broker to all, which is a highly advanced information operation in and of itself.

Canada has only limited technical intelligence capabilities but has, on occasion, been caught carrying out industrial espionage and other similar intelligence operations with national support. They seem to remain friends with everyone while continuing to act in their self interest in the information warfare arena. And even when they get caught with their hands in the cookie jar, they manage to walk away friends without apologies.

4.7 Israel

Israel seems to have a hard to comprehend ability to continue to exist while taking advantage of its closest allies in the information warfare arena. It spies on the US and gets caught again and again and yet the US continues to support it militarily, financially, and in every other way. It sells and has long sold military technology secrets to the Chinese, and did so even during the Cold war and when the US was very unfriendly toward China. The US barely reacted.

Israel also provides security technology like firewalls to the US military. And when that technology has unexplained code that seems to have the potential for exploitation, the US just continues to use it. They stole fighter aircraft and submarine technology information and sold it to China, were caught, and no punishment ensued. You might almost think that they are living under some lucky star.

> Israel's entries into the information security area demonstrate advanced technology capabilities as well as a clear understanding of the issues.

But intelligence operations are only the tip of the Israeli information warfare capability. In the battlefield arena, Israel was a key leader in developing increased tempo and accuracy of weapons and tactics that allowed them to survive attacks by their Arab neighbors in several wars intended to wipe them out. Their use of full spectrum information operations allows them to survive where they would have no chance of survival militarily.

But this apparent battlefield dominance does not make them perfect by any stretch of the imagination. They were completely snookered by Egypt in one major operation and seem to get caught more than most countries. Whether this is because they do more intelligence gathering than others in this arena or just don't do it as well is unclear.

Israel seems to have enormous weaknesses when it comes to propaganda and perception management. Their inability to have any level of control over the public discussions of terrorism is an example of their complete ineptitude in the public relations aspects of information operations. They get far less media coverage and they are generally treated negatively in the press across most of the world despite the fact that they are largely under attack from what is universally viewed as terrorism. Whether this stems from their own use of these tactics against the British in the creation of Israel, the nature of their treatment of Arabs as second class citizens, or simply poor public relations is unclear. What is clear is that none of these sorts of things have stopped other nations from controlling their perception around the world.

> Israel has some extreme weaknesses in information operations, particularly in the defense arena.

Israel also doesn't provide effective civil defense in the computer network attack arena. In fact they were even unable to continue their Internet-based military system of external communications effectively during a Palestinian attack on their Internet resources. Israeli companies have been attacked from Palestinian systems with impunity, and third party attacks against Israeli funding and political support in the US have apparently gone largely unanswered.

Israel demonstrates clearly that a nation state can have enormously effective military information warfare offense without being able to protect itself from attack while still maintaining enough of an edge to survive. But it also shows that the best defense is not a good offense when it comes to information operations. Their intelligence operations are outstanding, their offensive use of information in the battle field was a demonstration to the world in the 1960s and 1970s, and their ability to more or less individually target enemy leaders using advanced information technology clearly demonstrates an approach to information operations based on the powers of the state against individuals.

4.8 The India vs. Pakistan and China conflicts

In the ongoing conflict between India and Pakistan, information operations have been used by both sides and at levels of intensity warranting the term warfare. In fact, some might argue that the information warfare between these two countries has prevented them from escalating toward nuclear conflict. This is not just a war of words or a propaganda war, although those elements are most certainly present. This conflict involves direct computer network attack against vital elements of military and government infrastructure.

Well before the US was hit on September 11, 2001, information warfare efforts were coordinated with physical attacks. After the Pokhran explosive attacks in 2000, Pakistani attackers regularly attacked Web sites of Indian organizations. The sites of the Indian Prime Minister's Office, the Bhabha Atomic Research Center, the Ministry of Information Technology, and Videsh Sanchar Nigam were all broken into and defaced with anti-India obscenities. Groups like *Death to India*, *Kill India* and *G-Force Pakistan* openly circulated instructions for attacking Indian computers.

In December of 2000, the Rand Corporation warned that Osama bin Laden's Egyptian followers could immediately cripple the information infrastructures of Russia and India. At that time, bin Laden was still friendly with Pakistan, but there is nothing to suggest that these estimates were wrong or have been seriously diminished today. This fact was published in 2000 in India by Ravi Prassad, and of course the Rand corporation was studying the issue for the US government at that time.

After the attacks on September 11, 2001, Pakistan became allied with the US, at least to the extent that Musharraf could carry it off, and even in India some are calling for alliance between Pakistan and India to fight off China. In this case, the conflict between India and China comes into focus not only in the direct interchanges but in the more specific context of international relations with the US.

At some point, the US started accusing India of selling secrets about US warfighting capabilities and technologies related to warfighting to Iraq. The long-term effect was that the US decided not to sell military technologies to India that India wanted to have. Part of the response was an escalation of the conflict between China and India as reflected in the apparently accurate assertions in the Indian media that the US was selling the same weapons systems to China that China had been caught selling to Iraq. But China was not being punished by trade sanctions while India was. This is reflective of the complexity of information warfare and perception management in international conflict.

The India Pakistan conflict is not without its Indian attacks of course. In one interesting incident, someone attacked Pakistani Web sites claiming that they were from India and that India has the best "hackers" in the world. It included:

> *"Pakistan's official website has been hacked. My India is great,... "We are the best hackers in the world. This is a final warning for you not to play around,"*

It is sometimes hard to tell if these are kids playing around, mild skirmishes, gamesmanship, or part of more advanced capabilities. Perhaps it is a display intended to back off official information attacks, or perhaps it is intended to provoke them. And of course Pakistan must ask these questions of itself.

India and China have long been at conflict regarding sea lanes and information technology, a seemingly strange mix, but an interesting one to discuss... somewhere else. The sea lanes aside, there is a strong competition between India and China in the information technology arena, even if it does not get to the point of being a war. They both provide large-scale outsourcing of information technology to the world, they both have massive investment in education of their billion people, and they are both vying for supremacy in the outsourcing and soon primary creation of information technology. Will democracy or communism win?

4.9 The Israel vs. Palestine conflict

This information conflict is particularly interesting because it shows just how violent, how serious, and how dynamic information warfare can be while showing how closely it is integrated into the rest of conflict. The intensity of the ongoing conflict between Israel and Palestine – or whatever name you want to use for the set of peoples who call themselves Palestinians and the areas they occupy and govern – has a very substantial dynamic range and it changes rapidly.

In the late 1990s and early 2000s, the intense negotiations for peace after a cease fire was arranged caused, first, a significant lull in the hostilities, then, a raging open sort of warfare at every level. It did not reach Clausewitz's version of *all out war* in the sense that all resources of both groups were not seeking total annihilation of the other, but it did get to the point where bombings, ground, and air actions were taking place almost daily. And then, by the mid 2000's it reached a relative lull again. During this period of changing conflict intensity, there was a corresponding escalation and de-escalation in the information arena including a broad array of methods and types of information warfare. Some examples of the types of actions include:

- **Propaganda:** Both sides escalated internal propaganda campaigns surrounding the intifada.
- **Public relations:** Both sides, but more effectively and at larger scale, the Palestinians, created large-scale public relations campaigns and got the press involved in telling their story.
- **Direct computer network attack:** The Palestinians directly attacked Israeli companies and infrastructure via computer network attack.
- **Indirect computer network attack:** The Palestinians threatened companies doing business with Israel, including one incident where they attacked AT&T computers.
- **Funding and support:** Both sides, the Israelis to greater effect, escalated their use of computers to encourage

donations. In some cases Palestinians used computer break-ins to take credit card information that was then used to fraudulently charge donations.

- **Counter Funding and Support:** The Palestinians directly attacked US-based Jewish organizations via computer network and, in one case, succeeded in gaining access to a large Jewish donor and political supporter list.
- **Threats and intimidation:** Some of the information stolen by the Palestinians in one computer network attack was used to threaten and intimidate donors to Jewish causes.
- **Luring and killing at an individual level:** In one incident, a Palestinian woman used computers via online chat sessions to lure an Israeli teenager into a personal meeting where he was killed.
- **Political activism:** The Palestinians used computer sites to encourage and support political protests and similar events and to generate support by projecting false information.
- **Information in support of limited operations:** Both sides had limited operations under computer-mediated control. The Palestinians coordinated activities by providing training and instruction manuals on Internet-based computers so that attacks would be designed to optimize the media effects in order to intensify public relations surrounding their acts. The Israelis were trying to minimize the media presence in their operations in order to minimize the perception that they were warlike.
- **Command and control of military operations:** Both sides used computers in direct support of military operations.

All of these actions were escalated as the intensity of the conflict increased and deescalated as the conflict decreased. This represents one of the clearest examples of the tight linkage between information warfare and physical warfare and the escalation and deescalation process.

4.10 The China vs. US conflict

The conflict between China and the US has been going on since the end of World War 2. The information warfare aspects of it have always existed in one way or another. *Communist China*, *Red China*, *the Reds*, *I'd rather be dead than red*, and similar sayings were part of the strategic US propaganda campaign to keep the notion of conflict in the minds of the US citizens. China had much less of a need for a special propaganda campaign because they took control over all media directly and basically became a closed society.

The Viet Nam war represented a clear demonstration by the US that China would not be permitted to dominate that region of the world uncontested, and a clear demonstration by China of its willingness to go the distance to influence its region regardless of the US perception that it was ready to take over the world. These surrogate conflicts were carried out in many places and not just with China. In some sense these small conflicts that displaced direct conflict between *superpowers* were information conflicts in and of themselves. They were tests of political will and national commitment that brought out internal propagandists and dogmatic opinions that led to political changes in China and the US.

> The US decided to give away its global leadership in information technology just when it was emerging as a dominant technology.

As the information age raged in the late 20th century, the US held a dominant position up into the late 1980s, but for some reason that is unfathomable to most, they decided to not continue to be competitive against the Pacific Rim in computer hardware and against the subcontinent and Europe in software. So the US started to lose its edge by dismantling its educational system. This continues through to the modern day where the US has thinned its support for education to the point of near crisis and will soon lose any remaining dominance over the world because of its failure to invest in the strategic advantage of a better educated populace.

China was not so foolish. In fact, China had exactly the opposite situation in the late 1980s. They had a vastly undereducated population, nothing to speak of in the high technology arena, and little or no hope of much more than self sustainability in a highly agrarian society. But the Chinese leadership somehow managed to understand what the US leadership missed, and they started to invest heavily in building up their industrial economy and to educate an enormous number of people in technical engineering and scientific disciplines. While they had a severe deficit that took them more than 20 years to overcome, they sustained the effort, and overcome they did.

> China decided to invest heavily in the emerging information technology just as it started to build up steam.

This transformation of China, particularly the coastal areas in the Pacific Rim, combined with the enormous population and vast area, finally started to pay off in a big way in the late 1990s when China gained sufficient critical mass to become the precursor of the powerhouse it is in the 2005 time frame. It started to be able to offer its labor at low costs in areas beyond textiles, wood, manufactured goods, and other similar low-tech areas. In information technology and engineering, it started to be able to keep up with anything the West could offer in terms of quality while keeping prices low. And as the Internet prospered and China increased its capacity in the key areas underlying it, they came to be increasingly competitive with the US in the very technical areas where the US was declining.

At some point in the middle to late 1990s, China made the critical decision to go into direct information conflict with the US and other competitor states. It decided that its dominance over the long run depended on its ability to use information technology as a direct and indirect tool. In the direct sense, it started to go after building up specialized expertise in information technology as part of its advance in all scientific and engineering issues. It invested in people and grew a capability while restructuring its institutions both of government and of education to meet the emerging challenges of the information age. In some cases these were directed at general knowledge, but an intense effort was also directed at the application of information technology for information warfare, ranging across the whole spectrum. In the indirect aspects, China realized that global financial operations, infrastructure control, communications and coordination of military operations, supply and logistics, transportation, and command and control were directly operating and likely to continue to operate through the common information technology that was being put in place throughout the world. It decided to invest both in people and in things that would allow it to gain control over the life blood of the information technology over the long run.

> China's leadership understood the strategic issues of global dominance in the information age while the leadership in the US was seemingly bent on self destruction through its need to divide its people for political ends and missed completely anything of strategic value over time frames of more than one election cycle.

With this realization, China, determined not to be left behind, decided to take the lead. And it did so, with the full force of its growing economic engine and educational system behind it. While the West thought it was impossible and undesirable to prevent internal groups from forming and communicating, China set out to solve the hard engineering problems of internal tracking of everyone and their use of information technology. While the US thought information technology would be the downfall of communist society, China took the problem on and learned how to watch and control what people see and where they go in the information age.

At the emergence of the dominance of information in warfare, the US had the lead and applied it well in the first Gulf War. China learned the lesson well and started to engage in a determined effort to struggle against the attempted global dominance of the US. And so its doctrine of information warfare emerged.

China has been in more or less direct information warfare conflict with the US since the late 1990s. Only the US has failed to realize it. So as China takes strategic shot after strategic shot, the US sits on its hands thinking that this is just the nature of global economic pressure. As the average boat rises, some boats must fall, so as long as the wealthy in the US continue to be wealthy, the fact that the middle class will shrink, perhaps into oblivion, is not even disturbing to half of the leadership. And the other half cannot get enough of a fusion of ideas together to make the picture cohesive enough to win the votes needed to change the situation.

> It will take a generation to overcome the educational deficit created in the US over the last 25 years, and the only way to buy their way out of it will be to import the expertise from the very nations they are at information war with!

So the US, even though it has plenty of people who see the very things I am discussing, and even though it is engaged at some level in the eternal discussion of these issues, will not react in any meaningful way until a crisis emerges. And when the crisis of 25 years of failed national educational policy finally does emerge, it will take another 25 years before the change can be undone. The notion of throwing money at a problem works for most things in the US but there is only so much money, in the sense of real value, and unlike building buildings or buying books, educating minds takes time – lots of time. It will take a generation for the US to regain its intellectual capacity, unless it imports it from the rest of the world. But the very places to import this expertise from are the places at information war with the US. And because of the amount of human resource available in China compared to the US. It is very possible that the US will be unable to keep up in the strategic time frame even with maximum investment.

In some very real senses, China has already won the information war with the US and done so without firing a shot. They have done it by winning the strategic war for the hearts and minds of its own people over a 25 year period of time.

But in some other senses, the US is not done yet. And if history tells us anything it is that nation states don't often go softly into that good night. The US has opportunities, at least to be able to defend itself, and perhaps even to regain its global footing.

The US could push forth a serious educational reform process that includes large-scale re-education of the people educated in the last 25 years to provide them with some sort of decent living in the industrial and information age that is emerging. It could work feverishly to use the seed corn that remains from 25-50 years ago to educate the emerging generation. It could start to embrace science over religion in matters relating to research and development while still addressing its environmental concerns and those of the rest of the world. It could apply science to building the economy of environmentally sound survival and work toward self sufficiency instead of unlimited global interdependence. There are any number of other strategies that might be undertaken in concert to emerge again from the ashes. But the real question for the US is whether they can emerge at all from internal political disputes, finger pointing, pork barrel politics, perception management, self-deception, and financial irresponsibility, or whether collapse is inevitable.

And then what?

There have also certainly been times in the past when formerly dominating nations have been squeezed to the breaking point on economic or other issues and come out swinging. If China is not careful, the US will be pushed to act rashly. And history's examples are typically directed at warfare in the physical sense. It is not far out to imagine the propaganda campaign that leads the US into global war with China. First it's the trade advantages they are taking, then it's the conflict in the Taiwan strait, then its shooting down a plane or blowing up an embassy, and escalation happens. War is often used instead of good policy and strategic thinking when nation states get in trouble, and with only a few nuclear blasts, the US could eliminate the Chinese information warfare advantage. But then what would the world look like?

World War 3: We're losing it...

While on the subject of the China and US conflict, I think it is a good idea to provide more details. These are brought in through extracts from a paper written by one of my students, Benjamin M. Butchko as part of a report he wrote for one of my classes at the University of New Haven. This content is included here without further citation.

Chinese societal structure is tightly controlled by the government and communist party. With a lack of free market economy, the government exerts strong influence on Chinese corporations and economic areas. Beginning in the late 1970s, China began trying to make its economy more market driven, while maintaining the political control of the Communist Party. This has resulted in immense growth in the GDP, which has quadrupled since 1978 and provided a buffer from the Asian financial crisis of 1998 and 1999.

Chinese citizens are not highly restricted in their ability to enter the US. One of the primary drivers for this lack of restriction is the large economic market that China represents for US business. The Chinese government uses this to its advantage by employing scientists, students, business people, or bureaucrats, in addition to professional civilian and military intelligence operators as tools for information gathering. Today China has about 4,000 US businesses thought to be involved in these activities.

The PRC uses a variety of government, and government sponsored, organizations in information gathering and intelligence operations. Due to the socialistic and cultural institutions present in China, the PRC leverages its wide span of influence to engage multiple state sponsored institutions.

Chinese government is controlled by the Communist Party. The Party has a controlling position in the State Council as well as the People's Liberation Army (PLA). Through this combination, the Party exerts control over political, military, governmental, and commercial activities in all of China.

The State Council controls the PRC's military-industrial organizations through the State Commission of Science, Technology, and Industry for National Defense (COSTIND). This organization was created in 1982 and is responsible for the integration of civilian research, development and production efforts within the military arena. A multitude of interrelated industrial institutions have been created to meet the larger national goal.

The PRC has two intelligence organizations that engage in espionage efforts directed at the US: the Ministry of State Security (MSS) and the PLA General Staff's military Intelligence Department (MID). However, due to the widespread Chinese Government sponsored organizations, the foreign intelligence organizations account for a relatively small share of the information collection operations. The bulk of information gathering activities are directed by the State Commission of Science, Technology and industry for National Defense (COSTIND). Since the early 1990's, the PRC has been increasingly focused on acquiring US and foreign technology and equipment, including particularly dual-use technologies that can be integrated into the PRC's military and industrial bases.

COSTIND is responsible for implementation of the 863 and follow-on Super-863 programs. 863 was started in 1986 and aimed at narrowing the gap between the PRC and the West by the year 2000. This program was extended and expanded in 1996 with the Super-863 program. The program budget was split between military and commercial projects with main areas of program application including:

- Astronautics
- Information technology
- Laser technology
- Automation technology
- Energy technology
- New/exotic materials

These programs gather technology and assimilate them into military systems and industrial bases.

The 16-Character Policy was formalized in 1997 and holds that military development is the objective of general economic modernization. This provides for alignment in the economic and military goals of the nation, and as the economy expands, further funding is provided for military R&D, systems purchases, and increases in the military-industrial complex. Specific areas of developmental emphasis within this program are:

- Battlefield communications
- Reconnaissance
- Space-based weapons
- Mobile nuclear weapons
- Attach submarines
- Fighter aircraft
- Precision-guided weapons
- Training for rapid-reaction ground forces.

This plan acts as a guide for economic and military development throughout the Chinese bureaucracy and economic market.

The United States is vulnerable to both information gathering operations and information warfare attacks due in large part to the openness of the US economy. The prevalence of Chinese visitors and residents in the US offers the potential for a wide network of Chinese agent infiltrators. Economic incentives to access the enormous and emerging Chinese market support increased interaction and provide China with a large 'carrot' to entice and coerce US firms. A Rand study demonstrated that, because the US economy, society, and military rely increasingly on a high performance networked information infrastructure that presents attractive strategic targets for opponents who possess information warfare capabilities:

- The US use of the Internet and prevalence of US companies placing large volumes of information in this public area allows for easy access to intelligence gatherers.

- The US military complex has been shifting to greater use of commercial off-the-shelf technology for military systems. Thus, the line between commercial and military technology has blurred somewhat.
- The US society and economy is reliant upon the National Information Infrastructure. This infrastructure is designed, maintained and operated by corporations and individuals that do not place security at the same priority level as the military does. However, the military relies heavily on the commercial complex for logistics and information support, and thus would be greatly affected by a major commercial infrastructure disablement.
- US military systems are becoming more dependent on information technology to allow a smaller force to be more effective. While this provides great benefits, it creates vulnerabilities to information attacks.

Most of the losses of US technology to China occur in the form of commercial, scientific, and academic interactions between the US and the PRC. The bulk of information is gathered by various non-professionals, including PRC students, scientist, researchers, and other visitors to the West. Joint venture operations, commercial companies and fronts in Western countries, purchases of equipment, general foreign visits, and use of students also help this program. Professional scientific visits, delegations, and exchanges are utilized heavily as a means to gather sensitive technology. According to Rowan Scarborogh of the Washington Times, "Almost every PRC citizen allowed to go to the US as part of a delegation likely receives some type of collection requirement, according to official sources." This coupled with the multiple thousands of delegations that visit the US each year provides for an incredible number of espionage opportunities.

The PRC uses a set of strategies to gain intelligence form the US:

- Purchase or licensing of equipment and technology.
- Acquiring interest in US technology companies, providing an inside track to information and technology.

- Diversion of dual-use technology, leveraging technology that is licensed only for commercial technology transfer and applying such to military programs.
- Joint venture operations with US technology companies applying pressure on US commercial companies to transfer licensable technology in joint ventures.
- Use of front companies to illegally acquire technology and provide cover for intelligence operatives, and as a means to gain visitor access to the US.
- Illegally transferring US technology from third countries.
- Covert espionage (within and separate from Chinese intelligence organizations).
- Recruitment of ethnic Chinese in the US who have access to sensitive technology.
- Exploitation of contacts made by students educated in the US. In some cases, the students may be encouraged to study in the US and asked to not only gain a degree and information at a university, but also to intern at a US corporation and bring back information to China.
- Pressuring US firms to transfer technologies to the PRC as a prerequisite for Chinese market entry. Having an emerging market with one quarter of the world's population provides a great deal of leverage in making these sorts of demands.
- Covert espionage is claimed to the most heavily used method of technology acquisition. This is done through personnel from government ministries, commission, institutes, and military industries independently of the PRC intelligence services.

Methods have shifted from the primarily illegal and covert methods, which were historically employed, to a heavy reliance on overt and legal activity. Information targeted also includes business driven information targets such as commercial strategic management information, bid proposals, price structuring, and marketing plans.

Unsolicited requests for US defense industry science and technology program information are the most frequently reported activities by Chinese information gatherers. They use *headhunters* as a guise to solicit information from employees involved in S&T work. The Internet is utilized for access to company Web sites, bulletin boards, and postings. Attendance at conventions and seminars offer opportunities to collect information directly as well as opportunities to meet individuals that may later be contacted directly with further questions. Foreign employees working for US companies are targeted by foreign collectors. Cultural ties are leveraged to establish rapport.

The PRC has used joint ventures with the oil and gas business to increase their technology base. They have made contractual arrangements with Shell and Exxon for exploration at different times, and then let them operate in areas that looked promising, but had been determined to produce only dry holes. However, they managed to keep the details hidden from the foreign companies until they had already shared modern and sometimes proprietary technology. According to George Tenant, former CIA director:

> "*Many of the countries whose information warfare efforts we follow realize that in a conventional military confrontation against the US, they cannot prevail... These countries recognize that cyberattacks…against civilian computer systems in the US represent the kind of asymmetric option they will need to 'level the playing field' during an armed crisis against the United States.*"

Beijing has the world's largest program of information warfare development. In testimony before Congress in 1997, Michael Pillsbury detailed the widespread threat of Chinese efforts to gather information and develop asymmetric attack means. The key to asymmetric threats represent attacks on information. Mr. Chang Mengxiong, the former senior engineer of the Beijing Institute of System Engineering of COSTIND stressed that "*even if two adversaries are generally equal in weapons, unless the side having a weaker information capability is able to effectively weaken the information capability of the adversary, it has very little possibility of winning the war."* Thus, supporting the strategy of information warfare tactics to defend against or even offensively attack an identified more advanced adversary as the US. China is databasing "famous scientists" overseas, including home addresses and China visitation history. This provides for refined targeting of technology experts of interest. Classified material are obtained through personal relationships, bribes, or computer attack. Following the 1991 Gulf War, China initiated a full-scale campaign to develop its information warfare capability at strategic, operational and tactical levels as part of its overall military modernization effort.

Several large-scale investigations related to Chinese espionage efforts that have been undertaken and widely publicized. The allegations of Chinese influence in the 1996 presidential election, allegations that a Department of Energy National Laboratory employee passed secrets to China, and allegations that the Clinton administration made sensitive US technology available to the Chinese through the sale of US satellites and computers are good examples. These investigations have serious political overtones that make it difficult to identify truth from posturing However, confirmed losses of sensitive information demonstrates that espionage activities are underway and produce successful results.

Chinese integration with US companies has become widespread. Most of the biggest manufacturers of information infrastructure, operating systems, laptop computers, cellular telephones, printed circuit boards, radio frequency identification (RFID) tags, smart cards, and components are now either owned by or produced largely in China. From Cisco to IBM to Microsoft to Motorola and on down the line, US companies are rushing to feed at the Chinese trough as the Chinese are sucking the technology knowledge and engineering capabilities away from the US. Add in the recent arrests of Chinese espionage agents who leaked classified data from some of these companies and the picture is unmistakably clear. China is at Information war with the US.

4.11 Other pairings

Information warfare is not limited to the examples I have provided. It is always underway and always has been, to a lesser or greater extent, for all pairs of nation states. Every embassy in a foreign land is part of the information conflict between those nations and used both for escalation and deescalation. Diplomacy is information warfare by another name. Who are the best candidates for open information warfare between nation states? No doubt Iran and Israel are pretty directly in conflict today and the war of words is heating up. The dispute between an emerging Kurdistan from the remnants of Iraq and Turkey would be only the start of that conflict. The former elements of the Soviet Union are constantly shifting and have interesting capabilities for information warfare. For some reason, African and South American states have stayed out of the high technology conflict arena for a long time, but will they remain there?

4.12 Groups and coalitions

Alliances are forming and breaking all of the time, and these groups have their own elements of information warfare. In addition to the basic notions of communications and cooperation between allies, joint military operations take very tight controls and this integration of technical capability leads to access, insight, and the potential for very harsh results during conflict when sides change. Coalition warfighting without very strong technical capabilities brings the potential that battlefield situation awareness can be brought to the opposition. The potential for spies and leaks of tactical information increases dramatically with coalition fighting. And groups form at all levels of information conflict, from propaganda shared between Arab nations to the European Union and its attempt to unify views of historically diverse populations over common ground. As groups form and break up, their detailed knowledge of strategies and tactics, individuals and their roles, and structure of how information operations work become increasingly problematic leading to eternal change as the only hope for long-term survival. Clearly the potential exists for an enormous set of issues to arise in this context.

4.13 A roll-up discussion

The presence and nature of information warfare between nation states is clearly an issue for the world and most of its people. But I don't have the time or space to spend much more time on it because if I do so you will be completely bored out of your mind before we go onto the next subject. So in the way of a roll-up, I have decided to provide an extract from an article documenting an on-line discussion I had with Consulting Editor, Ravi Visvesvaraya Prasad and summarized by Ravi with more credit to me than I deserve. The article appeared in an airline magazine in December of 2005. Ravi is asking the questions (Q) and I am giving the answers (A). See if you can use linguistic analysis to tell when he helped out with the answers.

Q. Please can you analyze the latest developments in the fields of Information Warfare, Psychological Warfare, Media Manipulation and Disinformation.

A. Following the collapse of the Soviet Union and the end of the Cold War, attention turned away from pure battleground strategy and tactics, and more attention started to be paid to the concept of "Non-Violent Warfare", warfare without shedding blood. This comprises many facets – network and electronic warfare, economic warfare, cultural warfare, psychological warfare, technological warfare, political warfare, ideological warfare, perception warfare, spread of disinformation and manipulation of the media - both in the limited context of warfare between groups, and in the broader context of the perception of people across the globe.

Larry King, a US news and interview personality who has a daily show on the Cable News Network (CNN) has, for the last several years, been touting the phrase "Perception is Reality". This is a truism at best, but more importantly, it is a reflection of the move away from the use of science as a basis for differentiating fact from fiction.

For the people of the Third World, the great dangers associated with a highly skilled, powerful, rich, and manipulative group of people in control of information are great indeed. The decisions they make are based on the information they have. As that information is controlled, so are their opinions and their decisions.

Q. Much before Larry King, it is prescient that in the very infancy of radio broadcasting, in the 1920s-1930s, Mahatma Gandhi, on a visit to a British radio studio, remarked: "This indeed is Shakti (the Hindu goddess of power and strength).

A. Actually Mahatma Gandhi is a superb example of an information-age general. Over two thousand five hundred years ago, the Chinese military strategist, Sun Tzu, had said: "A great general is not one who vanquishes his enemy on the battlefield. A great general is one who persuades his enemy to drop his swords and flee from the battlefield without putting up a fight."

In the centuries-old Chinese military classic "The 36 Stratagems of War", there is a chapter which says that a great general is one who quietly and subtly persuades the enemy's leaders, and indeed the enemy kingdom's entire population, to adopt one's own aims and objectives as their own, one's own belief systems as their own. Another chapter says that the test of a great general is that he gets the enemy to welcome him into their fortress without even realizing that there is a war on.

I don't know if Mahatma Gandhi had ever read Sun Tzu or even heard of Sun Tzu, but he persuaded the British that it would be in their own best interests to walk out of India. Gandhi skilfully utilized all the media channels available to him to gain sympathy for his cause from his 'adversaries'. Gandhi's organizing of nationwide strikes is a superb example of economic warfare. Just as Sun Tzu said "Get the enemy to turn his strongest weapons on himself", Gandhi skilfully used the British media and legal systems.

By Sun Tzu's criteria, Mahatma Gandhi would be a greater military general than his contemporaries – Eisenhower, Patton, Rommel or Montgomery. Everything that a 21st century general should do, Gandhi has already done.

Q. There are contradictory trends in perception management today. You have "embedded journalists", Fox News, Al-Jazeera and CNN. The Internet enables everyone to instantaneously access news and all shades of opinions from all over the world. Satellite channels and DTH enable everyone to get all the news and views they want. But the Internet also permits anyone, anywhere to set up a newsletter or multimedia broadcast channel, with an instantaneous worldwide reach, to propagate whatever propaganda or ideology they want – hate speech, ethnic cleansing, terrorism, without any controls whatsoever.

A. To see the effects of these issues, one only has to look at Iran and Egypt and Turkey. There is no big ideological conflict underway between Iran and Egypt, but the differences they have over peace with Israel are as striking as can be. How is it that one Islamic nation can be at peace with Israel for three decades with no military conflict whatsoever and no threat of such a conflict, while another Islamic nation declares to the world that there can be no peace until Israel is dead and buried?

Even though Iran does not have a land border with Israel and has never historically been at war with Israel, since the overthrow of the Shah in 1979, successive Iranian governments have used tight control of their media to propagate their view of Israel as evil and Jews as Satanic. After three decades of this propaganda, Iran has enough mass of domestic opinion to be able to publicly declare a policy of genocide against Jews, without a single Iranian raising a voice in protest against the Iranian president's call for Israel to be wiped off the face of the earth.

Egypt also has tight control over the media. But after he signed the Peace Treaty with Israel, also in 1979, Anwar Sadat – and his successor Hosni Mubarak – used their tight control over the media to sell to the Egyptian people the advantages of peace with Israel. After the same three decades of perception management as in Iran, the Egyptian population has been able to forget the three wars with Israel in the past half-century, and indeed forget all about the three thousand year-old conflict between Egyptians and Jews. Today, the Egyptian man-in-the-street is friendly towards Israel and the Jews – all because of the message of peace propounded by the autocratic Egyptian leadership.

Turkey also has tight control over the media. But since the days of Kemal Ataturk, the generals who run Turkey have used strict censorship to suppress hate speech and to positively promote secularism.

In the United States we have been getting mixed messages, but the most coherent voice has been that of Lou Dobbs, another CNN news lead, who has, from the start of the US involvement in the "Global War/Conflict/Struggle on Terrorism/Radicalism" identified the conflicts underway in Iraq and Afghanistan as the "Global War on Radical Islamist Terrorists".

But it is not the nature of Islam as a religion to be terroristic or to hate anyone. The Egyptian and Turkish regimes have shown that it is not the Islamic religion that creates the conflicts, any more than the Spanish Inquisition with its Christian basis implies that all Christians are evil or warlike. Indeed, both the Egyptian and Turkish regimes have used the media to successfully propagate to their populations that violence and hatred in the name of Islam is totally unacceptable.

It is the manipulation of perception through psychological methods aimed at the population of a country by its leaders that creates hatred and warfare and allows the populations to be manipulated into conflict, war, and even genocide. The German people were subjected to this in the years leading up to World War II, and the people of Iran are being led to it today.

Q. For several years you have been cautioning the US defense and foreign policy establishments that they are in an information-age war with China, and don't even know it.

A. Yes, China is winning the information-age war against the United States, and most of the people here in the US don't even know that there is such a war going on. This war has been going on for at least ten years, and we in the US will wake up one day soon to find ourselves defeated. But the leadership of the US will no doubt say, just as they did immediately after 9/11, "But nobody could have ever imagined...". And most of the people here in the US will believe it because they are now so well-controlled and cowed.

China has very tight control over information to the point of even controlling at a detailed level what websites can be visited by anyone in China and who can use what methods on their computers. They surveil to an enormous extent and demand absolute obedience in the areas they control.

And yet, China has more people graduating from engineering schools at all levels by a factor of more than two than the US, more scientists better positioned in their careers, and a broader scope of scientific research with more funding than the United States. And the gap is widening quickly. The United States is moving toward worse schools, less support for education and science, less research money to the educational institutions that are the fuel for the future economy of the nation, dogmatic religious-based reduction or elimination of research that offends a particular religious viewpoint, and less money to support the rapidly shrinking human seed corn that is the professorate of modern universities. At the same time, China has moved toward more and better support for advanced education in the sciences, a broader spectrum of research in all areas, an increasing population of more highly-trained and skilled individuals, and no dogmatic cutting off of research lines with promise.

Since 9/11, the US has what amounts to an internal policy of declaring its own greatness and pointing to examples to keep its population believing it. There is an almost dogmatic pseudo-religion surrounding nationalism here in the US. The "Rally-Around-The-Flag" mentality has been exploited for dogmatic persuasion through media manipulation at all levels, by the use of money in the form of commercials, lobbying, and elimination of competitive forces, and of course the association of capitalism with goodness and freedom in general. Yes, we all know as if it were a fact that the Soviet Union collapsed under its own weight because the centrally controlled economic system couldn't manage it. But the US economic situation is not all that great. Five years of war, global competition, globalization, outsourcing, and offshoring have had serious negative impacts. Meanwhile, China with its enormous level of control over economic issues has flourished economically. So much for the dogma about the religion of capitalism.

Q: Has China won the information war without firing a shot in anger that could be traced back to them reliably? ...

India is also at risk from China. You may not have a repeat of 1962, but you just have to see how China has got the Myanmar junta regime in its grip to gain access to the Bay of Bengal, and the listening posts that they have set up in the Cocos Islands to monitor missile sites in Orissa and tri-series nuclear command in the Andamans.

And the Chinese are winning control over hardware and communications infrastructure. Just see how ZTE has bagged one of the largest telecoms orders in the world – from Bharat Sanchar Nigam. Just take a look at the Huawei facilities in Bangalore and the ZTE factory just outside Delhi.

Q. But Amartya Sen is of the opinion that India will be far ahead of China, mainly because the Indian media is free to highlight the shortcomings of the Indian political leadership. Amartya Sen constantly cites the millions of deaths due to famine in China that no one knew about because there was no free media in China to publicize to its leaders that famines had occurred.

A. As a researcher in mass psychology and its use in information warfare operations, I can tell you that things go awry when the leadership fails to do its tasks well and the media cannot show it to the people to allow them to change their situation. The well-being of the people is largely dictated by the quality of their leadership and the ability of their media to control the perceptions of their people. It is self-deception of the leaders that leads to these failures, manipulation of their own perception by their own choice of what media to watch and listen to and what they are willing to hear, group think of the political leadership, and so forth. But it only works when there is no outside objective opinion that is given adequate weight – as happened during Mao's Cultural Revolution.

In the world of the perception manager, perception is indeed reality, but over time, the realities of the world come to the fore, and even the most skilled perception manager cannot prevent starving people from wanting food or poor people from wanting clothes and housing. This is what Deng Xiao Peng realized which Mao ZeDong did not. And so Deng started the economic reform process in China, but retained strict censorship of the media.

Q. US think tanks have developed various computer simulation models of how the India – Pakistan relationship will evolve over the next few decades. How realistic are such war gaming simulations for a "non-violent" information-age war?

A. The war between India and Pakistan runs hot and cold, but in the information arena, it always is, always has been, and likely always will be a hot war.

What does the future hold for this information-age war? Whether it is outsourced work for the world in the field of information technology, creating an enormous engineering capacity from scratch in a single generation, advancing scientific research on a broad spectrum of issues, or creating an educational system capable of giving quality capabilities to a large and diverse population of people in the sciences and engineering, India has a clear lead in this conflict.

Over time, this will lead to increasing differentials in quality of life, understanding of issues, communications, transportation, manufacturing, and all other areas that ultimately lead to improvements in lifestyle, survival rates, and the ability to adapt to the changes coming in the world in the decades to come.

But neither India nor Pakistan are perfect societies, any more than China, the US, Egypt, or Iran are. India has a culture of castes, is a massive breeding ground for bribery and corruption, and has more starving and dying people in its massive and increasingly male-selected population than can be sustained over the long run.

Will India be able to overcome its enormous population problems and feed itself while still building its intellectual capital? Only time will tell.

Will Pakistan's military dictatorship be able to free it from its religious zealots by changing the culture and educational system to create a future for its people? The Turkish military generals on whom Pervez Musharraf models himself may have been able to do so, but in Pakistan's case, this is much more difficult.

About the influence of the religious zealots in Pakistan preaching hatred against India, as a researcher in the use of mass psychology in information warfare operations, I can tell you that no matter how much you and I might come to hate each other, our personal hatred is unlikely to escalate into actual warfare between nations, unless and until it is promulgated to larger populations of people and shaped in terms that make the larger groups act in malicious ways toward each other, even though there is no personal reason for the vast majority of them to do so. It is a sort of family bond, grown out of dogma of one sort or another, and manipulated through the media using dis-information, selective information, and methods of deception.

Q. In conclusion, do you see the forces of evil and violence making more effective use of new information and communication technologies than the proponents of peace and goodwill?

A. If history tells us anything, it is that with today's information technologies and tools, the future will be bleak indeed unless people with the power to stand up for independent truth do so before it is too late. Our future can be like the Spanish Inquisition or it can be like the visions we see in fiction. It is up to us to mold it.

It is ironic that in the state of Mahatma Gandhi's own birth, Gujarat, Narendra Modi has used the government media at his command to successfully convince the 90 percent of the population who are Hindus that their lives are in danger from the 10 percent of the population who are economically disadvantaged Muslims. Modi has used the latest mass psychology techniques to assert that genocide against Muslims is socially acceptable by portraying it as self-defense, and he has won election after election with this.

In contrast, the Turkish and Egyptian regimes, which are for all practical purposes, military dictatorships, have used the media and mass psychological techniques to promote peace.

But I am an optimist and I believe in the value of science and facts that can be demonstrated repeatedly by experiment and independent skeptical observation. I also believe in the value of a broad education and well-funded and heavily-supported University research and education as the path to a bright future. I think that the future of all nations resides in large part in their ability to create a large enough population of smart, well-educated, and non-dogmatic people who can overcome whatever adversities they face to advance the well-being of humanity as a whole.

But many of the key decision makers in positions of power around the world do not hold to these views and think that they should be the decision makers for humanity, carrying out their mandates by manipulation of the information that makes their cases for life and death, war and peace, and ultimately every aspect of the lives of the billions of people they seek to control.

A quick summary of information warfare between nation states is in order. Basically, these enormously well funded and very powerful group entities apply these resources in different amounts and with different strategic and tactical reasons against a wide array of problems they face and with a wide array of combined methods and approaches.

Information is a full spectrum element of warfighting between nation states. It is present as a means to ends, as a direct end into and of itself, as a fundamental element of intelligence, command and control, supply and logistics, support for the war at home, training and controlling the warfighters, battlefield information dominance, eliminating the fog of war and creating new sorts of fogs, and all other aspects of military operations. And this has been true since the time of Sun Tzu thousands of years ago.

But information warfare also extends far beyond the direct military operational roles it plays. It is core to the development of the people of these nation states and these people lie at the heart of their information capabilities as a society. Without the great societies underlying these nation states, their capacity for conflict dies. The hearts and minds of the people form the most core element of strategic information conflict and of national prosperity.

5 Criminals and information war

The criminals of the world have not missed the lessons that information technology has brought to the businesses of the world. At every level, criminal organizations have integrated information operations into their criminal enterprises and gained the advantages of scalability, decreased cost, increased flexibility, and increased tempo. Criminals have used the same capabilities that non-criminals use and brought crime to a whole new level through advances in technology and their ability to use it.

In a spoof on this topic for a recent conference, I started talking about how criminal organizations use information and information infrastructure, got increasingly enthused about what I was saying, and then started to launch into a great marketing pitch for criminal enterprises of all sizes to use my new and improved cybercrime capabilities to gain all of the advantages that the largest businesses have for half the cost. That's right – half off! Buy now and get a special bonus... sorry – I get a bit carried away when it comes to opportunities in new and emerging markets.

The basic advantages for criminal organizations to go to information technology are really quite strong. In the old days, you could only really do crimes against folks you could touch, using personality and looks or breaking and entering. Sure – the phone made things a lot easier, but it really didn't eliminate the problems associated with being caught. But in the information age, things have changed pretty dramatically for the better for the criminal organization. Here are just some of the benefits identified for criminal use of information technology:

- **Lower total cost of criminality:** Yes, the cost savings of using automation to run your criminal enterprise make factors of ten or more reduction in staff feasible, leading to far less exposure to insiders turning good or being pressured by the police.
- **Faster tempo means more crimes more quickly:** You can commit literally thousands of frauds per hour at the push of a button.

- **Wholesale crime replaces retail:** Because of the leverage of information technology, you no longer have to commit crimes one at a time. The same new rip-off scheme can be undertaken with hundreds of thousands, even millions of potential victims, all within the space of a few hours, and from all over the world. Turn your retail crime into a wholesale crime spree at the push of a button.
- **Sometimes you could end up in a legitimate business by mistake:** That's right, some things you might think are crimes in the physical world may turn into legitimate businesses if done oversees and using the information infrastructure instead of the mails. Take advantage of the lag in laws to gain legitimacy.
- **It's harder to catch you:** That's right! With information technology you can rapidly pass through hundreds of jurisdictions creating huge problems for law enforcement and their subpoenas. Have fun with your friends and cross the global infrastructure in seconds, putting up stopping points in foreign lands with different laws. Even use enemies of your country to commit your crimes, acting like you were their friends.
- **The punishments are lower:** Yes, It's true. White collar crime is punished to a far lesser extent than comparable blue collar crime. Steal a few thousand from a bank and you will get the FBI after you forever. Steal a few million from a company and they will cover it up. They might even pay you to tell them how you did it! Use the same methods against their competitor and do it all over again.
- **You can commit your crimes from anywhere:** Tired of that snowy city, that disgusting apartment, or living with your parents? Try going to the Caribbean! Lie out on the beach using your cell phone and laptop computer to commit crimes in the afternoon, party all night, and when you are done, do some crimes in Europe! All from the comfort of your seaside apartment.

- **Automation reduces the workload:** Tired of those dreary dialing sprees where you make hundreds of calls in a day and only hook a few lonely hearts? Go on line and leave the hard work to the computer. Send out millions of love letters in a few days for less than you would pay in phone bills for a week of dialing. So what if only one in a thousands falls for your scam. When you send out a million tries a day, it only takes a few free spirits to send you $100 each and you will be making it big with little or no overhead.
- **You can hit and run even faster:** Afraid of getting caught because you are a real nasty? Have no fear! Spend a day on crimes from anywhere in the world, then hit the road, filtering the money through your Cayman Islands bank accounts. Take a few days of vacation, then access your criminal enterprise from over the Internet from wherever you are. It's quick, easy, and efficient, and it works around the globe and around the clock. Hit and run so fast that only the CIA will be able to catch you, all at a low low cost that will amaze you.
- **Your criminal enterprise can practically run itself:** With new automated enterprise management software, you no longer have to be there to check up on everything. Want to watch every move they make? Use the new Web Cam service to record your hoods in action and critique them when you have time. Having bookkeeping problems? Use our automated bookkeeping software to track your assets, bets, on-line gambling, and even your cash income.
- **Stop worrying about taxes:** Afraid of the tax folks tying your money to you? Don't even think about it. With the new multi-hop money transfer and stock ownership systems, you can profit indirectly from your crimes, and remove any direct connection between you and your criminal enterprise. The IRS can't track the money when it goes through the diamond dealers and cash transaction systems sponsored by our friends at *The Base*. Al Capone never had it this good!

- **Take advantage of technology for better anonymity:** Have no fear, anonymizer is here. Brought to you complete with onion routing and cash per use cell phones, you can get direct access to information and operations data with practically no trace whatsoever. Never worry about wearing gloves again. Use your own cell phone and run through our remote access stations located in malls and coffee shops around the world. No muss, no fuss, complete anonymity without the hassles of taking special precautions.

With this as an introduction, and before you read the rest of this chapter, think for a bit about how easy it might be to commit crimes and run criminal organizations in the information age. Then think in terms of conflicts.

The US war on drugs is intimately linked to the US war on terrorism, therefore the US could use the same methods in fighting the drug dealers as the terrorists. Bring in the military and start searching all of the communications so as to can ferret out the terrorist drug smuggler criminal elements where they live. They have nothing to fear unless they are criminals. And who is more criminal than one of those child killers who supports abortion or those drug dealing doctors who prescribe to their patients. We will watch it all and only use it when we find it meets the level of terrorism – except of course when we happen to see a crime – or slip up and release all of the data to the public. It seems the slope is slippery indeed.

The basic premise of this book is that warfare is about conflicts, and one of the key conflicts that has been around since the beginning of time is taking things that other people earn without their permission. The war on drugs and the war on crime are prime examples of governments using the terms and notions of warfare when dealing with criminal acts. And some even try to escalate criminal acts to acts of war as part of their approach to removing the restrictions to what the government can do to their own people.

5.1 Classic criminal enterprises

Classic criminal enterprises include the Mafia, the Chinese Tongs, the Russian oligarchies, and other similar organizations. These are formal organizations that are lifelong membership groups existing based on some common features, typically including family relationships, racial similarity, common languages, and a hierarchical command structure. They are called classic in this context because they existed long before information technology became dominant, they were international, large-scale, involved global trade, and in many ways operated as other large global businesses operated before information technology came to the fore. They also have some other common themes.

Lisa Tortorice, a student in one of my University of New Haven classes did a report on organized crime as it relates to information technology. I am including selections of it here without further citation. It can be found in its entirety on the all.net Web site.

13 percent of computer-related crimes in Great Britain in 2000 were committed by organized crime groups.

According to a study published in 2001 by the Confederation of British Industry ("CBI"), 13% of the cybercrimes that occurred in the UK in 2000, were committed by organized crime. These organized criminals have long been known for their involvement in gambling, prostitution, drug trafficking, and their control over unions. One of the most obvious characteristics of organized crime is their threat and/or use of force. Making money to support their criminal enterprises is the main motive of organized crime, and with the additional money that can be made by committing computer crimes, organized crime groups have begun to make a transition into using new technologies. Some traditional crimes, such as money laundering translate very easily into this new environment, while others, such as drug trafficking do not. Credit card fraud and other Internet-based scams are a new source of income for these groups.

Organized crime groups are able to take advantage of the technologies that legitimate businesses are using. Correspondence between organized criminals can occur via email just as we all correspond via email. This causes their communications to occur immediately and from any place where there is a computer and Internet connection. With new encryption technologies they are also able to hide their tracks from law enforcement making it difficult and more expensive for officials to follow an information trail. A re-mailer, or multiple re-mailers, can be used to remove identifying headers from emails and replace them with false information.

> Organized criminals use the Internet just like the rest of us do, but they are in many cases more interested in the use of technical surveillance methods by police and how to counter them using cryptography and multiple hops and jurisdictions.

The same holds true for cell phone use and the criminals' ability to prevent their communications from being traced. Phone calls can be sent through multiple carriers, local phone companies, wireless, and satellite networks. Organized criminals can also clone cellular phones by copying an electronic serial number. Cloned phones cost the cellular phone industry millions of dollars a year and make it more difficult for law enforcement agencies to trace the calls with accuracy.

The Internet itself has become a tool by which organized criminals can commit their crimes. Money laundering may be easier and faster to do with the increased use of online banking systems. The transfer of money from one account to another can be completed by simply filling out an online form. This eliminates face-to-face interaction and can be done from and to anywhere in the world.

The anonymity that the Internet offers allows criminals to commit their crimes with less of a chance of being caught. Since these on-line banking systems and credit card systems, contain so much information about their customers, it is possible for organized crime groups to obtain bank account and credit card numbers by breaking into their systems. Organized crime groups are recruiting younger and more Internet savvy experts to join their groups, and there is also evidence of these groups buying this type of information from computer experts. Another means of defrauding Internet users is through websites which "sell" goods on-line. Customers purchase the product, but never see it, and organized criminals now have another credit card number.

The Internet has become a new means of distributing child pornography and has added a new dimension to the problem of sexual exploitation and the trade of women. People can sell, buy, and advertise child pornography and prostitution on-line through the use of images and streaming videos. Scanners, digital cameras, and digital video cameras are technologies that criminals employ to advertise their pornographic products on the Internet. Much of the child pornography on-line comes from the scanned images of pictures in magazines and films from the 1960s and 70s. These pornographic images of children are posted on on-line newsgroups.

> Much of the development of the Internet has been led by pornography and criminal enterprises. And this was true of the video technology and the film technology before them.

Through advertising, organized criminals are able to let customers book appointments and ask questions about the brothel and the children that they are interested in. Internet chats are also used as a way to "recruit" young men and women and children to run away and become prostitutes. Interactive DVDs are also being made. These DVDs allow the user to imagine that they are on a date or are engaging in sexual acts with the "actor" on their computer screen.

The Internet is also used as a means by which organized criminals can sell illicit drugs and pirated copies of software, as well as conduct illegal gambling. These crimes are becoming global in nature because the Internet does not follow traditional national boundaries. This is also a problem for law enforcement since cooperation between countries can be difficult because of differences in laws. Organized crime has taken advantage of the differences in many markets, especially in software piracy. It is easier for criminals to pirate content by stealing a CD-ROM and replicating it, packaging it in countries where the piracy laws are less rigid, and selling it into markets around the world.

> The Internet is not only being used to commit crimes and support criminal enterprises, it is also used to create now sorts of criminal enterprises and new business models for criminals.

Imagine a world where you can get online and purchase cocaine, a DVD of child porn, an illegal copy of software and then spend a couple of hours gambling illegally. Or perhaps it is easier to imagine getting an email promoting a new stock from what looks like a legitimate sender. After reading through some of the materials online you think this is a great opportunity to make some money and hand over a good portion of your savings to organized criminals. Some of these crimes have already occurred and some of these may occur more often in the near future.

This is how organized crime has started to move into the Internet and the information age. Is it low intensity conflict? It sure is. Does it ever get higher intensity? You bet it does. The communications of the Internet have been used to set up murders and assassinations by organized crime groups, and in some cases these have gotten to the level of multiple homicides in one location committed by groups of organized criminals at war with each other. While these are not nation states at war, they produce collateral damage, they cost countries and their people a great deal economically, and information warfare components play a significant role.

5.2 Gangs

Gangs and gang violence exist all over the world. From the small street gangs that roam together for personal protection to the increasingly consolidated multi-city and even multi-nation operations, information technology and the use of information in conflict has grown substantially over the recent years. Much of this section is taken from a paper written by another one of my University of New Haven students, Ben Bergersen, who focused on cyber gangs. His article is also available on all.net and is quoted liberally without further citation here.

Gangs are illegal associations of people that seek to perpetrate crime, typically on the streets. Cyber gangs are groups of people that congregate on-line to perpetrate illegal and unethical attacks on computer systems and networks for financial gain and enjoyment. The members can be geographically separated and often communicate solely over the Internet. Members of cyber gangs have some traits of computer attackers as well as those of traditional gangs.

The *Crips* and the *Bloods* are two of the most well known traditional street gangs in the United States. They have spread throughout California and have existed for well over twenty years. With the start of African American Los Angeles gangs in the 1920's street hoodlums started to gather, join forces, and become gangs. *Crips*, *Bloods*, *Latin Kings*, and other gangs cropped up in Chicago and Boston, and eventually have started to spread even into medium sized cities and some small towns.

Street gangs attempt to replace the family of young children. They seek to become surrogate parents. In turn, the street child receives love and attention. These children provide services, legal and illegal, as they grow up, in exchange for receiving a new home and family. The gang is typically supported through illegal means consisting of extortion, numbers running, gambling, protection rackets, and drug dealing. Drug sales are the most profitable of these incomes.

Just as a legitimate business has a customer base to protect so does a gang. Streets are marked off as gang turf through graffiti markings on buildings and roads. Graffiti writers are known as taggers, as they tag the land with spray paint and gang symbols. Once the customer base, willing or not is marked off, the gang members seek to protect their investment. Guns and dogs are purchased. Guards are posted. When a rival gang attempts to trespass, warnings are given, fights ensue, and sometimes deaths occur. These conflicts sometimes reach a frenzy and gang warfare is the term commonly used to describe the situation.

Identities are protected as well. Aliases and nicknames are used for members instead of proper names. Rarely will you hear of John C. Smith the gangbanger. Rather, you get names like *Lefty*, *Scarface*, *Bull*, *Shorty*, and *the Jeweler,* which are used to scare rivals and protect the true identity of the gang members.

There are typically several levels in a gang. The groupies are people who want to be in a gang and hang around hoping to become members. Probational members are new and need to prove themselves. Full members have proven their trustworthiness and prowess through illegal actions. Full members have tattoos. Life members have extensive tattoos and are normally in for at least five years. Everyone wears the colors of their gang in pride and to further advertise their organization. Bloods wear red handkerchiefs; and Crips wear blue handkerchiefs.

As gangs seek to replace nuclear families they use tribal associations and methods. These tribal methods extend today to the cyber world of the World Wide Web and other communications over the Internet. Gangs started out on the streets. A sense of belonging occurred, territory was fought over, and weapons used. As people moved away to other parts of the country the members could go inactive or join another gang. Switching gangs is seen as loathsome to many, but being a part of a greater whole is not. Dr. Theodore Walker, Jr. of the Southern Methodist University contends that gangs are tribal in nature. Tribes are a group of people in a social organization that share similar culture, ancestry and leadership. A subcomponent of a tribe is the family. This is understandable as gangs seek to replace the parents and families of children. Given this definition, Dr. Walker pulls together multiple writers that say that the Internet is a retribalizing medium:

> *"Back in 1979, building upon work by Alvin Toffler and Marshal McLuhan, Vine Deloria, Jr. of the Sioux nations predicted that electronic networking technologies would have a retribalizing influence upon the modern world. More recently, Derrick de Kerckhove, successor to McLuhan, described the Internet's world wide web as a powerful retribalizing influence."*

So, assuming that gangs are tribal in nature, and that the Internet is a retribalizing medium, then gangs will use the Internet. This is borne out with creation of the now defunct Glock3 web site in New Zealand. Glock3 was created in early January 1996 according to the Emergency Net News Service published in an article published on the Infowar web site. The New Zealand ISP removed the web site after extensive pressure. Afterwards allegedly a man attempting to capitalize on the notoriety created another www.glock3.com website. Looking at these sources proves that the World Wide Web can and is used for traditional gangs. The Pueblo Colorado Sheriff's Office web slide presentation said the original Glock3 site used to provide:

- A potential link for 40,000 gang members in 700 cities
- Exchanging ideas on improving drug sales
- The best gun to use to shoot rivals
- The best drugs to use in your spare time

Today traditional gangs communicate in Internet chat rooms, using instant messaging and electronic mail. These methods are harder for law enforcement to find or stop. These are non-permanent methods that appear to disappear as soon as the messages are sent.

5.3 Drug cartels

Drug cartels were far more dominant in the late 1990s than they are today. It's not that they stopped selling or that consumers stopped buying drugs that reduced their import. It was an increased presence of so many bigger threats. Nevertheless, drug cartels and large dealer networks have used information and information technology extensively in their war with law enforcement. And if you don't think it's a war, you haven't been watching very closely.

The war against drug cartels uses the highest technology methods available to military organizations, involves complex air, ground, and sea coordination, and involves spies and intelligence operations that most countries wish they had against their worst enemies.

The biggest drug cartel issues seem to come from Colombia and Afghanistan. Now, in case you thought that the US and its allies won the war with Afghanistan, you might have missed the fact that in order to win against the Taliban, the US allied with the local leaders who ran the local drug cartels. Yes, that's right, the drug cartels were chosen over the religious fanatics who they were at war with, and instead of destroying the poppy fields and using the enormous military resources in place to eliminate both of the major headaches that plague Western societies, only the religious folks who attacked and killed 3,500 people on September 11, 2001 were destroyed. The drug cartels killed 17,000 people in the US in 2000 alone according to the fact sheet provided with citation to Journal of the American Medical Association, Jan. 19, 2005, Vol. 293, No. 3, p. 298 (see http://www.drugwarfacts.org/causes.htm). Of course this includes all illicit drugs, but it is a good guess that a significant portion of those deaths were related to Afghani narcotics. So given that US deaths in all wars since 1990 are less than the annual deaths in the war on drugs, it is hard to characterize this conflict as anything less than a war. It is certainly a highly violent and high intensity experience for anyone caught up in it. And nobody is more violent and intense than those in the drug business.

All of this should convince you that the war on drugs is a real war and that the use of information in this war is every bit as important to information warfare as any other war. So when we look at the information war on drugs, the first thing we have to notice is the propaganda war. Somehow, marijuana, which by the way is more widely used than the other listed drugs and yet kills absolutely nobody according to the same sources, is treated as more dangerous than cocaine or heroine. Going back to the famous government sponsored propaganda movies, one of the most famous is about the terrors of marijuana. But even ignoring this bizarre conflict with tobacco, which kills 435,000 per year in the US, and alcohol, which kills 85,000 per year in the US, the propaganda campaign associated with the war on drugs manages to keep it a high priority in one sense, generates lots of funding to fight it, and ends up involving educational campaigns in schools and on television. Not that this is a bad thing, but when a government has to work this hard to fight a war that lasts for scores of years without resolution, it does not bode well for a favorable outcome.

The integration of drug cartels with revolutionary militant groups creates a combined information and warfare capability that nears that of a nation state.

Of course information is used by the drug cartels in the widest possible range of ways just as it is used by those who fight against them. The two major Colombian groups involved in drug smuggling are the FARC in the south and the ELN in the north.

The ELN attacks information infrastructure to harass the government. It is relatively small, working in the border region and often crossing into Costa Rica to avoid conflict with the Colombian military. The Revolutionary Armed Forces of Colombia (FARC) on the other hand, is a fairly substantial military organization with a well defined hierarchical leadership structure. I will focus in on the FARC as an example of how far a drug cartel integrated with a military revolutionary group can go down this road.

The FARC is the largest, oldest, most capable, and best equipped revolutionary group in the Western Hemisphere. They operate predominantly out of the southern half of Colombia and increasingly have presence and operations in Venezuela, Panama, Ecuador, and Brazil. They control most of the southern half of Colombia with some 16,000 FARC members. The Colombian government granted the FARC administrative control over substantial territory to facilitate peace negotiations.

> The FARC controls a large portion of Colombia and uses information technology heavily in military, drug, kidnapping, money laundering, extortion, funding, and other operations.

Various estimates of FARC financial resources place their annual income between $500M and $900M, with the majority deriving from drugs. They fund their activities largely by protecting narcotics traffickers. Estimates of the profits to terrorist groups from their involvement in narcotics range into the hundreds of millions of dollars. Their main business is selling protection to the traffickers and coca farmers by guarding coca fields, laboratories, and airstrips from attack. They also tax cocaine production, aircraft landings, and some river traffic.

The FARC also generates income by kidnapping Colombians and foreigners for ransom and extorting money from businesses and individuals in the Colombian countryside. Numerous cases of kidnapping have been documented, some for ransom, and others for political or psychological reasons. Wealthy business people and land owners are extorted by a 'vaccination', whereby payment to the FARC vaccinates them against kidnapping or murder.

The FARC is trading drugs for arms with criminal elements in Russia and some of the former Soviet bloc countries including Chechnya, the Ukraine, and Uzbekistan. During peace talks, the FARC displayed an array of weapons and has a secure arms pipeline which delivers vast quantities of assault rifles, heavy machine guns, small artillery, and rocket propelled grenades, which significantly alters the balance of power against the government.

According to a recent congressional report, in spite of increasing aid coming into Colombia to fight the illegal drug industry, areas of cocaine cultivation increased by 50% in 2-5 years, and 28% in 1998 alone. The FARC benefited from the collapse of the Medillin and Cali drug cartels so this may account for some of the growth. An estimated 75,000 farmers in the FARC territories grow coca. The typical farmer's income is estimated as high as $2500 per year growing coca versus $300 per year growing legal crops. Not all of the crop increase is attributable to the FARC, but their sharp increase in membership suggests that much of it is.

> The FARC has the information technology sophistication of a typical medium sized business.

Given the global availability of information resources, it seems likely that the FARC could afford to hire available talent, including ex-Soviet experts in information operations, and acquire specialized weapons and technologies. The use of coca crops for trading purposes may be an even more potent method of gaining access to otherwise illegal offensive information operations capability because of the inherently illegal nature of the whole business and the willingness of all parties to take great risks in exchange for great benefits.

According to Associated Press reports, the FARC has substantial insider information including access to government files on individuals who live and work in Colombia, and they can buy almost any information they desire. For example:

- They obtained a key archive that lists the owners of Bogotá real estate and property prices.
- Local FARC commanders pioneered the use of laptop computers for kidnappings, storing databases of potential victims. When they stop vehicles at rural roadblocks, they frequently check to see whether the occupants are listed.

- They keep lists of all soldiers and police, have the citizen identification numbers and full details of all of the members of the army and the police. If you get stopped at a roadblock, they find your citizen ID number in their computer and identify you from it along with your income, wealth, ownership of land, and occupation.
- They have a list of military pilots with home addresses.
- They use satellite communications to distribute their database to remote outposts.

> They have access to sensitive government databases and use them to identify and decide how to treat each person they encounter.

These examples are clearly indicative of insider access to sensitive government databases. They just as clearly indicate that the FARC exploits such insider access to great advantage. They apparently take and update the information regularly, in large volume, and in digital form. They then implement programs to analyze and reformat the data, and they implement telecommunications systems and databases to allow its efficient use over encrypted links to remote sites in real-time.

What is not clear is whether this access is the result of a sophisticated technology purchasing capability or more general insider access. If this is simply the result of wiretapping or purchases, it can be easily expanded into offensive exploitation through information corruption or denial of services. If this can be expanded, it may be realizable to an advanced level by the FARC.

It seems clear that the Colombian drug cartels are using substantial information technology for their operations, and it is just as clear that this technology capability now extends to the FARC. Traffickers make use of the latest technology to protect and advance their business. They keep in touch by using Internet chat rooms. Each part of the operation feeds its information on the day's sales and drug movements to a computer on a ship off the coast of Mexico so that if one computer were taken down, it would be harder to trace the rest of the network. They use encryption technology far beyond what law enforcement typically has the capacity to break quickly. One US official said it took some of their best computers 24 hours to crack a 30-second transmission by the traffickers, making the exercise pointless relative to the volume of transactions taking place. They also use cellular phone cloning, stealing numbers that were already assigned to legitimate users, using them for a short period of time, then moving on to new numbers.

The FARC uses up-to-date hardware and software, and the FARC commanders use wireless digital communications in the jungles. They use primarily portable computers, this use is widespread, and it is integral to the operations of the FARC. The FARC uses secure-channel satellite communications. Their insider access and ability to integrate information for different uses indicates a level of sophistication comparable to most medium-sized US corporations with augmented capabilities for encryption and computer security and an augmented offensive information operations capability. It does not indicate extreme sophistication or the ability to carry out advanced coordinated cyber attacks.

The FARC has and uses information technology for all facets of their operation.

The FARC apparently also uses computers for accounting and to manage their finances, which means that information operations against the FARC might be successful in destroying their ability to carry out many of their operations as well as in severely damage them financially.

According to the US drug czar in 2000, the FARC uses wiretap equipment as part of its military capability. They are also in touch with European groups and may try to leverage Russian cyber warfare technology which includes a wide range of offensive and defensive personnel and capabilities. While the FARC is only one example of a drug cartel / revolutionary group, it is clear that they use information as a weapon of war.

5.4 Criminal information exchanges

Criminal groups use computers for a lot of different things, but one of the largest uses is simple communication. Actually, it's not all that simple. It turns out that criminals need reasonably secure communications, and unlike most corporations and individuals, criminals are willing to suffer a high degree of overhead for privacy and anonymity. The information age has brought them some of the best tools they have ever had for this purpose, but their use of these methodologies is hardly driven by the Internet.

Criminal organizations have used cryptographic systems for a long time. Drug runners during prohibition in the US used relatively sophisticated cryptographic techniques to send messages without detection or readability by the police. They used flash paper and still do to be able to destroy information rapidly should the police enter a premises and they used concealment to a large extent. Many of the same techniques used by spies are used by ordinary criminals. So I guess you could call them leaders in their field.

With the emergence of the Internet, criminals found a far richer field for their covert and private communications. The combination of relative anonymity, outstanding encryption capabilities, multiple jurisdictions involved, and the ability to act from anywhere were among the most important reasons other than business efficiency that criminals flocked to the Internet for communications.

The most common tool used by criminal organizations for secure communications is cryptography, and the most widely used cryptographic system is called *Pretty Good Privacy*, or *PGP*. For many years, Phil Zimmerman, who created PGP, kept control over the source code, kept it open and available for all to see, and did a fine job of providing the best security he could. But eventually, corporate America took it over and turned it into a commercial product, which is when it started to have vulnerabilities. The hard core folks that want a secure version in open source so they can check on changes use the *GPG* open source version, as I do.

PGP (and GPG) have the advantage that they are good enough codes to make it quite hard for governments to break them, but they take some amount of effort to use. It took me 15 minutes to install on my Apple computer system and its use is completely automated. While corporate America won't take the time and effort to use encryption methods very well because most of the individual users don't see any value in being careful and they won't pay the overhead for corporate-wide solutions, the criminal organizations of the world enforce careful and uniform use because all of their workers see very direct value in its use.

Another important cryptographic tool for criminals is secure on line chat systems. Unfortunately for them, they haven't yet figured out how to use the best tools in this arena and are generally relegated to using second class tools which make their chat rooms readily accessible. If they read this and want to know the best tool, they won't find the name here. It would be shameless advertising to put the name of one of my own products in my book. It's only ego to call it the best of course, but the point remains that criminals are better than corporations at protecting secrets, but not as good as some governments.

The advantage of chat rooms and similar methods lies in their instantaneous nature and their lack of persistence. Most criminals believe that chats are not recorded on the servers they use and that the police therefore cannot capture copies for evidence unless they are listening in on the chats in real time. They are wrong of course, but let's not tell them that. The real advantage they do have is that they can be used by pre-arrangement for short bursts of communications from anywhere by anyone. So I can show up at a local library for 2 minutes and you can show up at a different one for the same 2 minutes anywhere else in the world, and we can get a 2 minute chat done in relative anonymity then walk away. Even if the police know who we are and are searching for us, they will only have 2 minutes to figure it out, find out where we are, get the local police notified, and arrive at the library to capture us. That's pretty hard to do in practice.

Email is also a major communications method for criminal groups, but it leaves a lot more residual data, so encryption is required. And of course increasingly mailing lists and blogs are used with steganographic encoding to conceal the meaning of messages posted in public areas.

As more and more secure communications media comes on-line, criminal organizations will continue to exploit these technologies to their advantage using increasingly clever and technologically advanced methods to communicate while concealing the nature of their work. And law enforcement around the globe faces ever increasing problems in meeting the criminal's use of technology in secure communications. This has been true for a long time, it is true today, and will continue to be true for the foreseeable future.

5.5 Computers to commit crimes

Computers are also used for direct criminal acts. These range from sending ransom messages using electronic communications, to computer network attack to facilitate ongoing exploitation, to theft of information from computers, to denial of services against the Internet interface or internal systems of companies.

Denial of services using distributed coordinated attacks became all the rage in the late 1990s when eBay became the first large enterprise target of such an attack, but the attacks actually originated earlier than that. I was one of the first targets of these attacks and was able to track down the perpetrators from my home even though they went through intermediaries all over the world. They launched the attack because I had insulted them by indicating that anyone who got caught committing a computer crime could not be that good and generally indicating that computer criminals were not very good as security consultants for that very reason. Tracking them down took about 24 hours for me to do as an individual, but that was not the hard part of stopping the attacks. Unlike today, when this happened to my Internet site, law enforcement and prosecutors were not able or willing to do anything about it.

Unlike today, these acts were not considered worthy of anybody in law enforcement. These computer criminals got away because nobody in government was willing to prosecute them. But lately, law enforcement has been getting better at tracking these folks down and prosecutors have been more willing to prosecute. For example, the US Secret Service recently got a guilty plea out of Anthony Scott Clark, 21, of Beaverton, Oregon, who has now admitted his guilt:

> "Mr. Clark and his accomplices accumulated approximately 20,000 "bots" by using a worm program that took advantage of a computer vulnerability in the Windows Operating System ... The "bots" were then directed to a password-protected Internet Relay Chat (IRC) server, where they connected, logged in, and waited for instructions. When instructed to do so by Mr. Clark and his accomplices, the "bots" launched [denial of service] attacks at computers or computer networks connected to the Internet. Mr. Clark personally commanded the "bots" to launch DDOS attacks on the nameserver for eBay.com. As a result of these commands, Mr. Clark intentionally impaired the infected computers and eBay.com."

Computer network attack is also very widely used for collecting credit card data. For example, in an attack on *Cardsystems*, a company that stored credit card information as a clearinghouse, tens of millions of credit card database entries were stolen in a single incident. This company didn't even try to react quickly to secure themselves after this incident and they were eventually put out of business when all of their customers walked away. They had ample opportunity to do something about it. They were contacted by the FBI when the crime was detected, and they were offered assistance by a reputable security company who could have saved them, but they refused to spend the money to get an outside opinion, failed to do a good job of handling the incident, and they are out of business.

The criminals not only got away with millions of credit card numbers and the ability to steal hundreds of millions of dollars by using them illicitly, they also put a billion dollar business out of business.

Computers are also attacked by criminal groups as part of their efforts to gain secure communications capabilities. In one incident in the late 1990s, what appeared to be a drug cartel gained large-scale entry into a whole series of companies through their internal networking infrastructure.

> Criminal groups directly attack computers not only for the information contained, but for extortion, for communications, and for exploitation against other sites.

These capabilities have been secured by computer criminals in many large enterprises, but they tend to use them for communications rather than for direct attack against these companies. In this particular case indications were that these networks were being set up so that the criminals could coordinate activities and share data without allowing the police to track them. The police would have to find out about it, track communications through several global corporations, and be able to do it in time to catch the criminals at the other end. This group figured it would not be within normal police capabilities, and they were right. They were detected by private individuals who worked the issues individually.

In most of the detected cases that I am aware of involving computers being used to commit crimes, no prosecution ever took place, the police were never contacted, and there was very little done about it other than getting the criminal activities off of the networks. Recent legal changes in the US have increased the reporting of select data theft, but the vast majority of these crimes still remain confidential because of the potential for damage to reputation for the company that reports them and the low probability of recovery in any meaningful way. Stop the attacks and move on with business. That is what businesses do.

5.6 Using computers to get to people

Computers are also widely used by criminals and criminal groups to exploit people. This is part of perception management on a one-on-one basis. Techniques of this sort are given names like *pharming* and *phishing* by security companies to allow the media to be able to talk about them and to gain publicity for the defenders and about the crimes. It's a propaganda approach to selling security.

The reason computers can be used to get to people is that people tend to trust computers, something that is unfathomable to those of us who work in computer security, but quite common among most people who are still apparently dazzled by the technology.

Computers readily produce excellent forgeries of all sorts, ranging from emails to Web sites. While most of them are easily ferreted out by looking at the details of the email in raw form, email systems are specifically designed to present emails in a format that is pleasing to the eye and, as a result, conceal the details that would differentiate a forgery from a real email.

> People are highly susceptible to the belief that computers are right.

But blaming technology and the writers of technology is not useful and it is not all their fault. The plain fact is that people are easily fooled and technology creates a barrier against the usual way people tell good from bad. People are used to dealing with other people more directly, and they are more easily fooled when the tell tale signs of someone they don't want to deal with are not present. But even with all of these tell tale signs, people are easily fooled. For more details on this issue, see my recent book "*Frauds, Spies, and Lies, and How to Defeat Them*".

The typical criminal enterprise will engage in large-scale forgeries of Web sites and emails. For example, they might send out emails that appear to be from your bank telling you that your account has been broken into. The solution is to click here on their secure server to reset your account information. The click will allow the attacker to collect the very data required to break into your bank account. This is not really a new approach. It was once commonly done by telephone calls and this will likely emerge again as people become more suspicious of emails. It is a well known elicitation technique used by all sorts of criminals and intelligence organizations.

Another common approach is to use the computer as a tool to misrepresent information. By creating a presentation to a user that causes them to act as if the presented information was true, many actions can be induced. This ranges from causing people to misoperate complex control systems, like aircraft and nuclear power plants, to making them dogmatically adhere to incorrect information, like records of debt or other similar financial records. Consider a criminal that breaks into a computer and creates database entries that cause the police to think that your address is really their address. The police will hunt for you, surveil your home and business, and possibly even arrest you because of the false computer records. And of course this takes the heat off of the criminals.

Criminal enterprises also fund computer attackers to break into computer systems for them and use the results to identify undercover operatives within their organizations. Finding an undercover Drug Enforcement Agency (DEA) agent, finding a witness in the witness protection program, finding key witnesses to crimes they have committed, and other similar efforts have led to assassinations of law enforcement agents and murder of key witnesses in criminal cases.

> Criminals use computers to get information on people for reasons ranging from theft to kidnapping to murder.

In even simpler examples, criminals with access to on line calendars and similar information are better able to break into and burgle homes, steal art works, and so forth.

5.7 Computers for fencing and laundering

Money laundering is the process by which illegally procured (*dirty*) money is turned into more safely usable (*clean*) money. In order for dirty money to be used in legitimate businesses it must appear to have been gained in a legal manner. In addition, dirty money can be used as evidence against criminals in an investigation and is often a target of seizure by law enforcement. Money laundering is a three-step process:

- First, the dirty money is placed somewhere. For example the money may be placed in a bank account under a false name, or kept in a safe deposit box.
- Second, the money is *layered*. Usually it is transferred from one account to another, maybe across national boundaries, to make it difficult for law enforcement to follow the trail of the money.
- Finally, the money is *integrated*. In this stage the money is made to appear as if it was received through legitimate business. There are many ways to do this, and they are documented in some depth in "*Frauds, Spies, and Lies, and How to Defeat Them*". Some of the techniques used are to create fake invoices for merchandise sold from one country to another, to buy and sell property and make the money appear to be obtained from the sale of the property, mixing the dirty money with clean money in a business bank account, mixing monies in stocks and bonds, or trading the money for a fungible commodity like diamonds or gold.

Money laundering has been committed by organized crime for decades. Most of these groups deal with a lot of money, and seizure of this money by law enforcement officials is a major concern since it can be used as evidence against them during trial and, more obviously, because they want their stolen money! While laundering a few thousands dollars is not all that hard to do unnoticed, doing this for tens or hundreds of millions of dollars is far more complicated and risky.

Many banks are doing business online and this has made it easier for criminals to transfer their monies from one account to another, even across national boundaries. Money laundering has thus become a serious international problem. The layering (through online banking) and integrating (through online auctions, gambling, etc.) portions of the laundering process can be done more quickly and effectively online, and the criminals can also take advantage of the anonymity that the Internet can provide to them.

Laundering is not only done by criminal enterprises of course. Many government and law enforcement agencies also have to create ways of getting money to run covert operations, and they use many of the same tricks. When Oliver North did money laundering during the Reagan administration, he used cash associated with sales to Iran to fund wars in central America. But far more sophisticated operations are used in law enforcement, often based on the use of seized assets from drug arrests. These assets are processed using false identities created using the power of the government, and complete backgrounds on individuals are created, sometimes as complete fantasies, but more often and more convincingly, based on facts.

Governments use much the same methods as criminals for money laundering in order to support covert operations on foreign soil. As criminal organizations innovate the governments that pursue them exploit their efforts. One wonders whether patenting these techniques and licensing them for government use might be a more profitable enterprise for the criminals.

The fictitious people then take the seized assets and sell them, use cash, and so forth to create clean versions of the same assets, which they then use to get involved with criminals again so as to catch them. Or in some cases, they use this money to buy new police equipment, support other operational needs, or whatever seems appropriate to those in charge.

5.8 Piracy

According to a study conducted by *Microsoft*, software piracy creates a $12 billion loss for the software industry per year. They also claim that more than 80% of all business software created is pirated in one form or another. In November, 2001, 31,000 copies of pirated software were found in Los Angeles. If sold at retail price they would be worth approximately $100 million. The software itself was made to appear legitimate, from the packaging down to the authentication codes. Among the products found were Windows 2000, NT and Millennium operating systems, Office 2000 and Norton AntiVirus. The authorities believed that the software originated in Taiwan and would have been sent to small software stores and sold over the Internet.

> Software piracy certainly costs billions of dollars in profits per year to the software industry, but these profits may or may not be reflected in job losses.

In December, 2002, a group of people who were pirating software across the US were arrested during a set of 100 coordinated raids. The criminals called themselves, *DrinkorDie*. The biggest difference between this group and other organized crime groups is that they claim to have committed their crime, not for the money, but for fun. Although not all of the evidence points to *DrinkorDie* being a typical organized crime group, they were still divided into groups of people with different tasks. In *DrinkorDie's* piracy efforts and in other organized efforts, the original suppliers are often insiders at a software company. They steal the software, *programmers* strip the software of protections, *testers* make sure that the unprotected versions of the software work properly, and *packers* divide the programs into small files and distribute them. According to a report of the Business Software Alliance, in the US alone 118,000 jobs and $5.7 billion in wages were lost due to software piracy in 2002. But this is a deception in and of itself. Additional profits to software companies usually lead to increased shareholder value, dividends, and executive bonuses rather than increases in jobs.

In addition to software piracy, organized crime has become involved in counterfeiting DVDs. Criminals have been operating freely in countries that do not have strict copyright laws. This allows them to have more of a chance of evading the law. Countries such as Russia and Malaysia have factories in which their counterfeiting takes place. Jack Valenti, president and CEO of the *Motion Picture Association of America* testified before a *House Judiciary subcommittee* stating, "large, violent, highly organized criminal groups are getting rich from the theft of America's copyrighted products. Only when governments around the world effectively bring to bear the full powers of the state against these criminals, can we expect to make progress."

Piracy of software and entertainment is very large because these industries have very high margins and very large volume businesses. This makes it inexpensive and easy to manufacture forgeries in large volume and distribute them with high profitability. If the margins were lower, the pirates could not afford to take these risks. Of course they don't have the cost of property development that the manufacturers have, so they don't need to recoup large initial investments in order to start profiting.

> Software piracy is very tightly linked to information warfare strategies at national levels.

Piracy also goes to the issue of designs associated with other intellectual property. Chinese manufacturers ignore patents while implementing their own versions of smart cards and RFID tags in enormous volume. They take copies of almost anything else you can name in the technology arena. This is one motivating factor for companies that try to move into China. They take advantage of the lucrative market in exchange for assurances from the Chinese government that they will have some protection of their intellectual property rights. It is a very lucrative extortion by the Chinese government as part of their information warfare strategy. Other governments participate in, or at least don't stop similar activities in their nations, making piracy of this sort a very direct form of information warfare.

5.9 Credit card fraud and identity theft

One of the emerging trends in organized crime is the use of credit card fraud and identity theft to make money, either by stealing it from the victim or through extortion of the companies that own the information. Organized crime has begun to either recruit younger "Internet generation" criminals or to buy credit card information from technical attackers. In June, 2002 computers at Arizona State University and several other colleges were attacked through the use of keystroke recording software. The software was used to steal the student passwords and credit card numbers, and to break into student email accounts. It is believed that these crimes were committed by Russian organized crime groups.

> Credit card fraud is a substantial source of income for criminals and they organize around Internet-based collection of the data used in these frauds. Note that Russian organized crime involves many former KGB agents.

In December, 1999 police in Toronto arrested 38 members of a Russian organized crime group. Although the group was based in Toronto, they had operations on at least four other continents across the globe. Allegedly the criminals gained access to credit-card information while it was being transmitted from stores to banks and also broke into ATM machines to obtain encrypted information. They were also able to create fake credit cards, and in using these stolen credit card and debit card numbers they were able to steal millions of dollars from banks and credit card customers. It is believed that the leader of the group hired someone to build equipment which could be use to intercept the information. In addition, they may have had the help of an insider to install the software on ATM machines and the credit card machines in retail stores.

In March, 2001 the FBI announced that more than one million credit card numbers may be in the hands of Eastern European organized crime. These numbers were obtained when extortionists attacked approximately 40 computer systems in 20 states in the US and then offered to help fix the vulnerabilities in the Microsoft Windows NT operating system that allowed them to break into the secure computers. It is believed that many of the credit card numbers that were stolen were sold to organized crime groups.

Identity theft is also becoming a major issue in the United States. With the increase of commerce being conducted through the Internet, stolen credit card and social security numbers, and other personal information has become extremely valuable.

> Identity theft is quite often mischaracterized as the same thing as credit card fraud. They are very different things. Identity theft involves taking over an identity and carrying out a wide array of things as if you were the individual. Credit card fraud is often simply a case of using someone else's credit card information to steal.

It is estimated, based on a fallacy that credit card theft is the same as identity theft, that in 2002 identity theft cost victims and financial institutions more than $700 million. According to this fallacy, identity theft cases comprised 43% of all fraud-related crimes reported to the federal government in 2002. Organized criminals have discovered that credit card frauds are a lucrative business and are paying insiders or attackers for the information required to carry out these crimes. But identity theft is typically carried out by individuals who are far more willing to take risks in order to use the new identity for all manner of other crimes. They also often walk away from the new identity after a time, leaving the original victim under a wide range of serious charges, ranging from assaults and robberies to home and car purchases, leases for spaces, and so forth.

Victims of credit card frauds can lose up to $50 maximum if they report erroneous billings or theft of their cards, while debit card victims stand to lose everything in their bank account. Identity theft victims can suffer far deeper cuts and have been falsely arrested, have ended up taking years to restore their financial situation, and have been treated badly in the meanwhile.

5.10 On line gambling

Traditional organized crime has been involved in gambling for at least 100 years. With the advance of technology illegal online gambling has become a major financial aspect of criminal operations. Gambling is a legitimate entertainment business in many jurisdictions, however, when the gambling is not fair and the odds are rigged by the house, or when states don't license the gambling and laws prohibit unlicensed gambling, it becomes a criminal act.

> The jurisdictional issues with on line gambling are far from solved and ultimately seem to come down to the ability of jurisdictions to win wars against each other in order to enforce their will.

Of course if I run a gambling operation from somewhere that gambling is legal and the people in the place where it is illegal decide to break the law to use my gambling site, why should I care about it at all? This jurisdictional problem is really severe in the networked world. Is it reasonable for someone who has a business that some radical remote government declares to be illegal to suffer the consequences of legal actions against what is legal for it to do where it is in business? Do I as the businessperson have to determine where you as a customer come from before doing business with you? Suppose you lie to me about it and use an anomymizing or pseudonimyzing service where the acts are legal? Am I responsible for hunting you down before I can carry out my $5 transaction?

The answer is, of course, force majur. If the jurisdiction has the power to carry its law to wherever I am, then it can force me out of business, and if it does not it can not. So in the extreme case, nation states enter other nation states, arrest people (also known as kidnapping in the jurisdiction where they do it) and bring them back for prosecution. Might makes right... or at least might allows you to take away other peoples' rights. And that is what happens in on line gambling.

But these issues aside, there are some more fundamental issues associated with on line gambling that are beyond my capacity to understand. Sure, betting on a sporting event can make some sort of sense on line. But who would play games of chance using on line forums when the house has complete control over what happens to their money? How hard is it for me to write a computer program that prints out:

- How much do you wanna bet? (get their money)
- Press here to roll the dice (simulate them betting)
- You crapped out (tell them they lose and take the money)

It's really easy to do this. And of course if I want to be more subtle about it I can produce a mix of rolls and make sure that every time I take $50 from you, you win back $46 of it. I can make it so you are never ahead but always close to catching up , until you bet big and lose big, and then I can help you get some of it back before taking still more. How do you enforce regulation of fair practices over a computer on the Internet in some foreign jurisdiction?

Gambling operations range from individuals running off-track betting operations to large-scale gambling houses with full casino capabilities. Card playing is increasingly popular with fads changing from game to game over the years. This also brings in side businesses like books and manuals, training, and even personal lessons in how to play the games.

There are two key issues to note here:

- If you are doing something you know is illegal, you are extortable. And most of the organizations that break the law are more than willing to take more money from your credit card than you thought, threaten you with extortion, or report you to the local police for violating local laws.
- Criminals commit crimes, against you if you get near them. They have used keystroke recording software to get credit card numbers for frauds, have stolen all the money in bank accounts when people provided debit cards, and extorted money from people who broke the law.

5.11 Child pornography and prostitution

In October, 2002, 80 people were arrested throughout Europe for their involvement in the trafficking of illegal immigrants. The criminals were arrested in Russia, Belarus, Poland, Spain, Portugal, Ukraine, Germany, France, and Austria. It is believed that the immigrants were being sold and forced into prostitution or slavery. These arrests were linked to Russian and Italian organized crime groups.

Organized crime groups try to commit international crimes in locations with regulations that are less strict. The cooperation between Russian organized crime and the Russian government has become a major problem in fighting the dissemination of pornography on the Internet. Moreover, these types of crimes are not prosecuted to the extent in Russia that they would be in other countries, making convictions all the more difficult.

> When discussing organized crime in the context of information warfare, don't forget that many members of Russian crime syndicates are also ex-KGB operatives.

It is estimated that 20,000 people are brought to the United States each year to be used as "sex slaves" and about 900,000 children and women are kidnapped or tricked into becoming sex slaves around the world. In 2003, a new US law made it illegal for US citizens to have sex with children while traveling abroad and for people to come to the US for the purpose of "sex tourism involving children" (sex tourism involving adults is legal). The profits made in the sex slave industry continue to provide money for other organized criminal activities.

From an information warfare standpoint, the pornography and sex trades are very important because they provide the means to extort money and information from people by exploiting their weaknesses. They are also a path for moving intelligence operatives from place to place. What better way is there to operate under cover?

5.12 The aura of legitimacy

Most big time criminals seek the aura of legitimacy, if only to cover up their seemingly unexplainable financial wealth in light of no discernible source of income. Because of the increased vigilance of tax collectors, it is becoming harder and harder to hide income if the revenuers come looking for you. Thus the richer the criminals get the more they seek legitimacy.

As they grow, criminal enterprises start to buy into legitimate businesses, using them to launder money, to conceal income, to facilitate their crimes, and to provide more opportunities for exploitation.

- **Buying into legitimate businesses** is an old technique for spreading criminal income out. The ideal business is a cash business which can support lots of benefits to its officers or other members, all as tax deductions. Another ideal business is the religious business, which is non-profit and thus not subject to taxes. More on this later.
- **Laundering money** is a necessity for many criminal enterprises that have large cash flows that become hard to manage and impossible to hide. Even if hidden, cash eventually gets replaced by governments, so burying it is not a viable solution. As a result, cash businesses are ideal for laundering money.
- **Concealing income** can be facilitated by any number of bookkeeping tricks explained in more detail in my recent book *Frauds, Spies, and Lies.*
- **Facilitating other crimes** through legitimate businesses often involves using the capabilities of those businesses for nefarious reasons. For example, by buying into a credit card processing business, it becomes far easier to steal credit card information. The information security business is ideal for the information warrior of today.

Clearly the criminal gains substantial advantages by taking over legitimate businesses.

5.13 The future criminal enterprise

If the past is a good predictor of the future, criminal enterprises will grow ever stronger and more efficient with time, leveraging information technology to the maximum extent feasible to optimize their operations, reduce workers, obfuscate their operations, avoid the police, move resources rapidly, take advantage of victims, seek out new victims, and communicate rapidly and securely. All of these are of course expected, but what unexpected things might criminal enterprises use computers for?

> Many of the largest companies in the world are criminal enterprises.

Some speculate that criminal enterprises will undergo large-scale consolidations and hostile takeovers facilitated by the increasing capacity to manage larger numbers of operations using smaller numbers of people by leveraging information technology. I think that this will happen to some extent but that crime is often self-limiting in this way. Unlike legitimate business, criminal enterprises depend on secrecy for success and secrecy gets harder with size.

Many criminal enterprises have legitimate business connections. Whether they are banking connections that facilitate money transfers or holding companies to cover up large numbers of transactions each taking small amounts of money, or links to government officials who send large volumes of business to their friends in a tit for tat arrangement, criminals are increasingly placing themselves in charge of large enterprises of all sorts.

I think that the future criminal enterprise is the better version of Enron. It is Microsoft that was found guilty of monopolistic practices and is now selling China the means to control news and propaganda. And Direct TV just paid $5M for violating the “Do Not Call” laws of the US. All of the better versions of Enron are still in business, and the fines they paid are only a small part of the money they made by breaking these laws. They do the analysis and determine which laws to violate. It's a business optimization.

Bill Gates was fined $800,000 in 2004 alone for violating antitrust laws by buying stock without adequate notice. But it's only a civil fine and no jail time is involved. This time he got caught, but how many times has he gotten away with similar things?

Of course many of these folks are small time compared to the well placed no-bid contracts awarded to Vice President Cheney's former and likely future employer, Haliburton. They get billion dollar contracts on a moment's notice with no bids or competition involved because the Vice President of the country used to be the President of their company. And they are later found guilty of overcharging for the work done under the contract, making much of it simple theft.

The Abramoff influence peddling scandal shows how corruption of the political system through lobbyists leads to massive thefts of monies from businesses trying to get equity in markets as well as a not so subtle form of bribery and large-scale theft of money from the people of the country. But it also shows how small numbers of criminals can subvert the legitimate governments of the world and destroy the political process. It is, in essence, a form of bloodless, subtle revolution carried out by criminals who are in fact information warriors. This will be discussed in greater detail later.

> When criminals start to use money to subvert the political process, they become traitors and seed revolutions. These sorts of conflicts can be understood in terms of information warfare and they represent the likely brightest future for the most successful criminal enterprises and their leadership.

How is this information warfare? When groups of wealthy individuals commit criminal acts to subvert governments around the world, through deceptions of the people of those nations and their legal systems, how can it be called anything else?

6 Information war and religious groups

If ever there was a historic source of information warfare, it is in the guise of religion. More people have been needlessly and horribly slaughtered in the name of religion than any other cause. And while the name of religion and the various names of God have been defamed by these outrageous acts throughout history, their influence on modern wars and conflicts at all levels is more pervasive today than it ever was in the past.

In some sense, religion can be thought of as nothing more than information warfare. While religion has helped many people come to deeper understandings of themselves and the world around them, and even helped many come to more peaceful ways, it has also been exploited by corrupt leadership to take advantage of and perpetuate ignorance, to fight wars for the gain of the elite and powerful, to kill millions under the guise of doing good, and to cause innumerable humans to suffer horrible lives and deaths.

Religion may be thought of as a computer virus of the human mind.

One reason the US Constitution was such a breakthrough was separation of church and state, and it was pretty effective for much of its first few hundred years. But today, the US has fallen prey to religious forces that threaten to destroy this critical principle and turn it into the same sort of society it proclaims to be so wrong in the Middle East. The major religious information war of our time, Islam vs. Christianity, has heated up to an intensity level comparable to the most vicious nationalistic wars.

Religion has been characterized as one of the clearest examples of a mental virus. Like a complex computer virus that exploits the fertile field of the computer's information processing capability to reproduce, a set of religious beliefs are information that reproduces in the information processing capability of the human mind. Like a computer virus spreads through computer communications, the mental virus of religion spreads through human communications.

World War 3: We're losing it...

In today's world, Christianity is the dominant religion, and Islam is a growing second. It only makes sense in the war of ideas that these two religions must clash in their struggle for dominance of the limited resource of human mental space and time. The information war in this context is quite literally for the hearts and minds, but it extends to body counts from time to time.

From the time of birth, teachings lead to belief systems that guide, inform, limit, and contextualize everything that people see, feel, and understand. Belief systems are very powerful contextual limiters on thought processes. By bringing people to believe in one thing or another they can dogmatically be led to kill others, eat their flesh, justify burning people alive, and commit horrors of all sorts. In fact without the creation of a belief system of this sort, it is unlikely that any sort of warfare can be conducted at a substantial scale.

> Religion is iwar in pure form.

Information warfare in religion starts just after birth when babies are ceremoniously mutilated, dipped in fluids, forced to eat things, and treated in any number of other ways that would seem irrational to an outside observer who had never seen these practices. In an attempt to explain these things long traditions are generated that include initiation ceremonies similar in many ways to the things done in college fraternities. Food, drink, song, movement, and everything else is associated with the religion, and its experiences become part of the daily life and ceremonial times of the members. This goes literally from the womb to the tomb and beyond as the members of these groups are taught different things about life and death and brought under control of their leadership by promises of eternal bliss or damnation.

Who prospers in this environment? The leaders of these groups, who often express privately that their belief often falters. Some of them end up abusing their child members, presumably for lack of a sensible love life, while others take the best of everything from the society to support their better lifestyle than the average participant.

6.1 The Christians

Christians may be historically considered in terms of the Crusades, a period in which wealthy leadership sent out hordes of religious believers to kill off those who didn't pay homage to their version of God. The Crusades were used to solidify power, take land, eliminate dissenters, and enrich religious leaders, but it never seems to have achieved any of its publicly identified goals.

After the Crusades, the world encountered localized religious versions like the Spanish Inquisition in which people who did not buy into official doctrine were tortured to force them to agree to whatever the current leadership demanded. Many were killed for no particularly good reason, but the survival of other religions demanded that they strike back with equal vigor, producing hordes of people who would rather die than yield to another person's belief. Torture until death or admission of crimes is a common religious theme. After all, you are saving them from eternity in hell and damnation, so a little bit of torture here on Earth is completely justified.

> Only those providing the money or labor are really fooled.

The Christians are represented most poorly in the US by the modern versions of Marjo, the televangelists. These television preachers speak of how holy they and their efforts are, more often than not, taking large donations from their television audience to build what amounts to palaces for their leadership to use. Giving is of course critical to the success of any religion because it allows them to do good works. Much of the donated money does get well used to proselytize. This is a process by which natives of all lands get to eat by agreeing to the religious terms of their benefactors. While many go along with these demands and act like they believe, only those providing the money are deceived about the belief systems. Neither the folks in the field nor their leaders nor those they subjugate buy into the momentary success of trading food for declaration of belief.

But in the long run, many of these campaigns are successful at converting the next generation to the religion and as these efforts succeed, more and more people in the area join the religion of their own free will. Religion is a transgenerational form of information war, and thus its strong support of human lifecycle issues. Besides, nothing compares to death to bring out those donations.

Christianity is also well known these days for its gay priests who end up raping children in their parishes. This has not been fully explored for its root causes, but who can deny that there is a link between the requirement of celibacy and the increased rate of child exploitation for homosexual sex? This notion that the priests must deny themselves Earthly pleasure is, of course, purely a human construct, instrumented by the direct link of the Pope with God.

> Most Christians are unknowing participants in the corruption of their leadership and much of their leadership is not corrupt. But any time religious dogma starts to dominate a society, the result is inevitable.

I have no intention of linking all Christians with crimes against humanity or with the Catholic branch of it. Most Christians, as most others in most religions are just normal people trying to live in peace, but when leadership ends up corrupt and dogmatic, war and suffering inevitably follow. Only the true believer or sociopath would be willing to undergo a lifetime of religious dogma to gain and maintain power and the benefits of power afforded to religious leadership. And of course as the largest religious group in the world, Christianity could be said to have the clearest right to dictate policy and viewpoints, assuming you are willing to allow anyone or anything do force you into doing and thinking their way.

The religious right wing in the US has taken a strategic approach to political dominance that violates the principle of separation of Church and State, brings about great divisions, and creates enormous social strife. This form of information warfare is driving ever growing military conflicts, carried out in the name of God.

6.2 The Radical Islamists

If the radical Christians are harsh, the radical Islamists are wildly more so. The most radical of the Christian religious groups are tame in their willingness to harm others for their cause compared to the current groups of radical Islamists. These fanatics misread and intentionally misinterpret their own religious dogma to turn what is normally a peaceful religion into suicidal maniacs willing to kill themselves and anyone around them in order to bring about the changes that their leaders claim to be in their best interest. All the time, many in their leadership are benefiting and living very well off of their suffering. The most radical terrorist groups are exceptions to this rule, and many of their leaders are true believers, but plenty are not.

> The same methods used by Christianity for dominating the future mindset of uninitiated people are also used by Islam.

Islam seems to dominate the landscape of today in the historically Arab lands and in conditions of relative squalor such as US prisons. It is perhaps most telling that by targeting the downtrodden of society, ideas can be brought to bear as weapons of conversion. Weapons that promise joy, peace, and prosperity forever are quite appealing to those unable to find their next meal, especially if it comes with a little food and water. Allah brought you this food and water and brought you me to help teach you the way. If this sounds very much like the Christian approach described earlier, welcome to my world. Prostelitizing to the poor in exchange for food works no matter what religion you want to put forth. Building houses, digging wells, teaching people how to survive, and generally bringing up their standard of living from squalor to survivable works for Islam as it does for Christianity as it does for the US foreign policy apparatus.

The teachings of the imams in religious schools that dominate the educational system of key countries in the Middle East demonstrate the religious information warfare approach to indoctrination as well as any you can find. Their inherent hatred for

Israel and all Jews is one of the themes of their teachings and they integrate general hatred for all things Western along the way. After all, we are violating God's law assuming as they do that the imam talks to God and knows how to interpret God's will.

The Islamists seem to have a far more serious problem with radicalism because of the general condition of their membership. While in Turkey, Islamic rule is democratic and in most ways reasonable and rational with respect to other religions and points of view, most other Islamic nations are far more radicalized. The strongest correlation seems to be their financial well being. Poor people living in squalor support radical religious ideas and are willing to die more readily for their beliefs because they believe they have so little to lose. Meanwhile, these same countries have forms of government supported by Western nations that dominate their populations, have all of the money, food, housing, and other forms of health and wealth, and take advantage of their citizens.

> Islam prospers today by bringing in the downtrodden of the World where most of the population is living in squalor or can be convinced they are oppressed.

The information warfare of Islam is largely involved in the different interpretations of Islamic law and the Koran. In this way it is again the same as Christianity. Christianity takes a single phrase in the bible which has at best unclear interpretation and translates it into a prohibition on homosexuality, while Islam takes similar phrases from the Koran and uses them to justify suicide bombers. This willingness of religious leadership to take a document, assert it was written by God, pick a single phrase out of context, misinterpret it, and turn it into violent or abhorrent action by members is typical of religious information warfare, a bane of humanity, and a fundamental problem that has to be addressed in all religions. Radical Islamists will have riots over a cartoon of Muhammad in Belgium but themselves promote both pictures of Muhammad in a favorable light and far more outrageous cartoons about others in their media all the time.

6.3 The Jews and Israel

For some reason that nobody can get around, the region of the World surrounding the city of Jerusalem became the center of Judaism (the first and second temples), Islam (the dome of the rock), and Christianity (part of the area where Jesus lived and died). For this reason, there is and will likely continue to be major religious dispute over ownership and control of that area for the foreseeable future. Perhaps the only real solution to this issue is the nuclear annihilation of the area, but this is not likely to be supported by anyone who wants to pursue peace.

> Isaac and Ismael were brothers. And that explains it all.

This information-based conflict has led to some of the most ferocious wars we have seen in the world as well as the indirect abuse of religion, rumor, and dogma to drive people in favor or opposition to one side or another. The Jews committed various bombings and similar terrorist acts to gain possession of Israel from the British after World War 2, while the Palestinians are trying the same thing today to gain possession from Israel. The Christians only seem to want visitation rights and not direct ownership. But in addition to these direct approaches, indirect approaches have also been used. For example, the Jews have been systematically stigmatized in Christianity as Christ killers for centuries and as back room dominators of Western civilizations since the late 1800s through a series of false conspiracy theories and supporting falsified documents. The Arab nations have historically supported only the latter theory, but have a blood feud that nobody can clearly define or claim to understand.

The best I can do is to say that it started when Abraham had two sons, Isaac and Ismael. Isaac ended up a founding father of Judaism while Ismael ended up a founding father of Islam. So I figure it's all about sibling rivalry taken to an extreme, but most modern folks seem to think it started again in the 1400s. At any rate, any dispute that goes on that long is likely to be unresolvable in this book.

6.4 The religious terrorist groups

This information-based religious conflict has gone on and on and been codified in religious doctrine, day-to-day teachings, phrases in languages, political speeches, arguments, individual exchanges at every level of ferocity, murders, group violence, wars, mass murders, and even attempted genocide. Whatever the real disputes may be, they are not the underlying issue of the conflicts surrounding Israel and never have been. These are religious wars.

And whenever religion goes to war, the extremists form terrorist groups. Whether their goals are forced conversion or simply death to the sacrilegious heathens (the other folks). These come in all religious varieties. Terrorist groups cited by the US State department that are religious in nature range from the radical Islamic groups to the radical Jewish Defense League to the Aum Shinrikyo (they have since changed their name), to the Bagwan whose group tried to use biological warfare to take over Oregon politically, to the group that committed suicide when the last comet came by the Earth in order to join the space travelers that they thought were using the comet as concealment for their starship.

> If you claim to be in a religion, you are. And if you kill folks over it, you are a terrorist. They are information wars founded on beliefs.

I know that those in the so-called legitimate religions will call these groups illegitimate, but they claim to be religions and as far as I can tell, they have the same right to make these claims as anyone else who can gather enough believers to make the newspapers and radio shows. Whether it's the Scientologists or those of the Bahai faith, they count in my book. So when anyone, even Charles Manson, claims to be acting in the name of God or gods, gets a bunch of converts, and kills a bunch of folks, they make it to the religious terrorist book from my view.

The key thing to note is that they use religion in the extreme to bring about rationalized violence against others. And this is information warfare in one of its most pure forms.

6.5 The Scientists

No discussion of religious information war can be complete without the Scientists coming into the fray. I capitalize Scientists here because I am not talking about the practice of science as undertaken by so many well meaning people, but rather the religion of Science in which we see dogmatic obedience to the principles of science as if they were a religious belief, and of course the treatment of science as Science by religious groups threatened by the truths of science.

> Religious dogma used as information warfare to limit science reduces the capacity of a nation to prosper but secures control over the nation by the religion.

In the conflict between religion and leap-of-faith-free systems of thought, the religious groups always seek to limit science while embracing it. This is true even in the most radical religious views. Even in religions that claim to be opposed to all but the most natural way of living in the world, science in name or in deed is used to survive and thrive. If they grow crops, they use the scientific method in their optimization of the growth of those crops by planting and harvesting during the right seasons. This is because nobody can really survive in the world by being completely irrational or ignoring the realities they see in front of them. Even members of the *Flat Earth Society* believe that things fall when not held up.

But at the same time, religious zealots seek to limit science wherever it could potentially interfere with their beliefs or get close to disproving some fundamental assumption they have made about the nature of God and the Universe. The debate in the US over stem cell research is a classic example of a society forcing itself out of a market and area of advancement based on religious beliefs of a vocal minority presented and amplified through information operations to create a dogmatic response by constituents resulting in delays in the advancement of science and technology. While the US does this, other nations push forward and develop the innovations that will bring this technology to the fore in the future.

That is not to say that stem cell research or any other specific area is either important or will make a future difference. Science is a risky business in terms of payoff for any particular area of long-term research. Most lines ultimately turn out to be inferior and more so as science progresses further. But without the investment, none of the lines prosper and, if you happen to choose poorly, you end up destroying the long-term capacity of your country to prosper. And of course a proper balance must be struck in terms of the safety of the population. Research in some areas may produce diseases or other indirect effects that are ultimately devastating. But we know that a failure to adequately pursue research dooms a society to collapse because history shows again and again that nations that fail to invest in technology and science lose their footing and become second or third class nations over strategic time frames.

> The dogmatic pseudo-scientist can be just as dangerous to the rest of us as the dogmatic religious leader. Either breed of information warfare hurts us all.

Fairness demands that I also rally against dogmatic Scientists. They treat science as if it were religion, and are often more damaging than their religious counterparts. In the name of science human experiments were undertaken in World War 2 by the Germans (where they were treated as war crimes) and during and after the war in the US (where they were kept secret for many years). Scientists who think that science is infallible or fail to do the necessary independent validation, reviews, and testing required for science to work, are abusing science just as religious fanatics who take a historic phrase out of context are abusing religion. Not all scientists are honest and people are not perfect, and that is why the scientific method demands independent validation and testability.

The use of public relations in science and the use of marketing techniques to turn science into a popularity contest is damaging. And the use of the powers of the state to suppress scientific views based on religious views of constituents and politicians is just as damaging.

6.6 The end of religion as we know it?

If science is right and religion is wrong, then over time the countries that apply science over religion will win in the market place. Or so the economic theory goes. But as smart as economists are, they have never been able to adequately codify human interactions and the effects of information on decisions. Against their own economic interests, people boycott stores with lower prices based on dogmatic adherence to religious views. They ignore obvious lies and deceptions and even forgive people who have stolen from them in large quantity and repeatedly based on a good sob story. *The devil made me do* it has revolutionized religion.

> Religion may have serious problems surviving the marketing upheavals associated with modern information technology. Will religion adapt and, if so, how?

As information technology and the information that psychology has gleaned about human frailty progress, more and more confidence artists have moved into the religion business. No taxes, less inspection, and unlimited access to parishioners who bring money in exchange for sympathy and want forgiveness which God will give without recompense as long as you support the church. If this view sounds overly pessimistic, forgive me. There have always been criminals who sought refuge in Churches, but today they seem to be aimed at refuge from the law and aimed at higher callings within religious institutions. Those that claim to be able to see into hearts are far more common than those who properly predict the outcomes. And as religious institutions push into the political arena, a backlash is eventually likely to happen. When will religions that push political views end up losing their tax exemption? Who can tell? But the abuse of religion, just as the abuse of corporations, both by criminals and those that should be seen as criminals by any other standard, is leading to a revolution against religion at the same time as religions are going stronger than they have ever gone.

The sales and marketing function of religion was long a combination of force, political influence, and fear of death and disease. Shamans of all sorts put themselves forth as the intermediary between the Earth, Wind, Fire, or other natural or supernatural source of life and death to gain and retain control over the people. But as times have changed, the marketing function of religions has advanced with the other aspects of society and science. The same basic principles apply, but there is an increasing need to create artifices that allow science to be successful while limited by God.

> The increasing use of information technology and larger-scale threats of information release lead religions to seek stronger intellectual property protection in the modern information society.

While religion uses multi-level marketing approaches, there is an increasing need to more directly defeat the attacks on Godly intervention by science and technology through an increasing use of more and more clever marketing techniques. In order to retain forceful dominance, information has to be seriously limited, so that only the high level priests can have access to the secrets of the religion. Scientology is the modern example of choice in this arena. In Scientology, there are secrets revealed to increasingly smaller members of the group as they progress up the leadership. These secrets are kept in select locations and protected by technological measures. Insiders have used computers to leak select portions of this information and been challenged in the courts under trade secret and copyright laws.

Will religion survive the Internet and the information age, where their secrets can be revealed and their attempts to defeat science as a rival lead to longer and more prosperous lives for those who apply science and technology? Will religion adapt to suppress science or move into adopting science as its own, describing it as the way of God reflected in the things of man? Will religion adapt as it always has by co-opting the world around it? Only time will tell.

7 Corporate information war

Most economic theories ignore the effects of practical monopolies and their ability to influence politics to allow their crimes to be ignored or inadequately dealt with. If you steal $500 from a bank, you will be hunted down by the FBI and put in jail for many years. If you steal $500M from the many shareholders of a public company and get caught, you will pay a few tens of millions of dollars in fines, high legal fees, and spend at most a few years in prison. If you deal drugs in large quantity and turn others in you will get a minimum sentence, but if you are at the low end of the totem pole, you will likely spend ten years in jail for buying what they sold you.

> Corporations participate in more direct and more clearly information based conflict than almost any other entity you can find.

So the rich dominate the poor and the large dominate the small, as it has always been. But what about the conflict between the largest of the large. The global corporations that have more resources than all but a few of the largest governments, and the ability to directly rule them from the top? How do they compete in the information arena? And does this extend to the lower levels and smaller businesses? Do niches disappear, get eaten up, or survive to fight another day? Information warfare in the corporate world is perhaps one of the most fascinating areas to investigate because it is the closest thing to absolute power and hierarchical society that continues to exist today, and warfare in the information arena is largely unrestricted by laws or other limiting factors.

In the domain of unlimited pure information warfare, corporations are as close as it gets. They fight each other directly with wars of words, living or dying largely by image and brand. And even beyond the perception issues, they depend increasingly on information technology for their survival and operational efficiency. When these mechanisms fail, companies quite literally go out of business, sometimes overnight.

And within enterprises, internal struggles for money and power dominate the success of individuals, despite the general notion that the most efficient results yield the most successful outcomes for the participants. Economic efficiency just flat fails in this context. People get fired because other people don't like them, or because of rumors or innuendo, or because they are too good and other people don't like the competition. And even when this is not true, people think that it is true, if only to explain to themselves why they were fired.

> Despite all attempts to systematize internal decisions in enterprises, judgment and perception still largely rule the day. Presentation and the ability to make and keep good internal relationships are key to influence and success in the corporate world.

All of these factors are information factors that may have little or nothing to do with job performance. And of course trying to find good measurements of job performance has always been a problem. In the end, sales are relatively easy to measure because the numbers of dollars showing up lead to the answer, but actual dollars are only a small part of the computation. For example, sales may be high and even dollars good, but the jobs generated may be lower profit or harder to carry out, or may detract from the overall business focus. The inability of companies to measure what they do leads to the need for other evaluative factors in the place of raw numbers.

And even raw numbers are poor indicators of success. While failing companies can always try to find some number of import, the tactical search for good short-term numbers every quarter also leads to a lack of long-term strategic investment that leads enterprises down the same patch as societies that fail to invest in the long term. Corporate information warfare is vicious, continuous, internal as well as external, and ever present.

7.1 Corporate espionage

Corporate espionage typically involves a very direct application of information attack techniques against competitors. This is a core area for the work of information protection, a field where I have worked for a very long time, and where the companies that fail to do an adequate job fail rapidly.

> Cardsystems went from dominance to death in two years because it could not survive corporate espionage.

Nobody claims to yet know who attacked Cardsystems, one of the worlds' largest credit card processors. But the FBI contacted Cardsystems because credit card information they held was being exploited to steal money from others. The FBI offered to provide contacts for Cardsystems to be able to rapidly address their problems, but Cardsystems ignored the advice, believing they could handle the problem internally. They are now out of business because they were unable to either defend against the attack or properly manage the image problems it yielded, the technical issues they needed to address, or the many other implications of this attack. I know this because the company the FBI referred them to was one of mine. Cardsystems refused all attempts at help and it ultimately killed them.

Compare this incident to the outcomes of other companies and you will see that many companies have similar incidents, some of the same magnitude, but few of them fail. The reason they stay in business is not clear, but they have at least one thing in common. They manage perceptions effectively. More on that later.

Corporate espionage is increasingly a major problem in the world of the information enabled enterprise. There is no realistic choice for most companies but to use information technology widely and to enormous effect. Without the business efficiencies brought about by this technology it is essentially impossible to be competitive.

The very operation of an enterprise is an exercise in information operations that is a battle between competitors for business efficiency. But yielding effectiveness for efficiency is almost always a failing strategy. When efficiencies overcome the need to be effective in the light of highly competitive environments, those who pursue this course fail. For example, it is more efficient in the short run to have everything centralized, thus the trend toward consolidated data centers. But in the long run, any individual site will fail at some point, and when it does, the aggregation of risk will quite literally kill a business. Ask any business whose sole major information operations were in the *World Trade Center* when it was hit. Most of them never reopened their doors. This terrorist attack was quite intentionally an information attack against businesses who had offended the sensibilities of a few radical Islamists.

And this is not an accident or novel notion. The *Irish Republican Army* (IRA) spent years directly attacking financial institutions by bombing their data centers. But the IRA typically bombed these facilities when nobody was present so as to minimize the loss of human life. They anticipated an eventual peace.

> One of the most successful and devastating acts of corporate espionage is the theft of business information by highly placed insiders and its exploitation by or to create a competitor. Every year, companies around the world lose half of their potential in a very short time from such attacks.

The range of corporate espionage certainly includes malicious attack to destroy a company, either physically or informationally, but not all espionage is directed at this. The theft of trade secrets, gaining patent rights without doing the work required to get them, damage to reputation, creating business opportunities, and gaining access to competitive advantage in bids and customers are also widely exploited advantages. And top on my list of most devastating and common attacks are insiders with access taking what they need to start competitive companies.

7.2 Perception management

I mentioned that success in light of large-scale successful attack is often the result of a good job of perception management. This is about managing the expectations of customers and business partners, and is an absolutely critical success factor for enterprises. When incidents happen, public relations and communications, technical, legal, and management response are required. If a single reason that Cardsystems failed could be identified, it would almost certainly be inadequate comprehensive response to the incident. And they are not alone. Citibank lost something like $400,000 in an attack in the early 1990s involving a Russian, an insider, and a few others. This was of no noticeable effect on the multi-billion dollar giant, but its public relations effect was massive, resulting in substantial lost opportunity and customer base for a period of years. But by now, industry has learned how to respond to the public relations effects of these sorts of attacks and this sort of attack rarely yields much harm because of the ability to manage perceptions and address these problems in a well defined and very publicly clarified manner.

> Managing the perception of any enterprise is fundamental to its success.

Of course perception management, manging the public and private face of an enterprise, is core to success of every enterprise. In the information war that is competitive business, getting the word out on your products and services, overcoming objections, preparing effective presentations, building brand, and securing a place in the market are fundamental to success. Without these, no business can grow unless it is by back room maneuvers with political cronies, bribes, payoffs, and other similar activities. And even these involve the same sort of activity, except of course that the perception is quite private and directed toward small numbers of select individuals with the power, influence, and access to resources to get the job done.

Weathering storms and presenting the best face on problems is part of the overall perception plan for any substantial enterprise.

This requires people skilled in information operations of this sort. If mishandled, these sorts of responses lead to exaggeration of problems and, in some cases, an ever growing pattern of negative public relations.

Lies are, of course, not desirable in any case. Except in politics, lies are not expected, and long-term loss of trust is severe when lies are put forth. People forgive mistakes in most cases, but lies are simply unacceptable and, in many cases, illegal. Ask Martha Stewart if you don't believe me. The only thing she did that was illegal was to lie to federal investigators.

In between lies and truth are presentation in the most favorable light, deception that is not an outright lie, avoidance, and misdirection.

- **Presentation in the most favorable light:** This method involves emphasis on the positives, minimization of the negatives, and ordering the presentation to keep the favorable on the mind of the audience.
- **Deception:** This is more direct in the sense that it is not simply presenting all sides in a skewed manner. It fundamentally involves trying to generate a different view than reality. As such it tends to be resented, but does not have to be a direct falsehood.
- **Avoidance:** In many cases questions can be avoided. If done skillfully the audience may not even notice it. If done poorly, you will be recalled on the question repeatedly and the audience may assume you were covering something up, which presumably you were.
- **Misdirection:** This technique involves directing people down a path that will lead them to your desired perception even if it is not what they would have thought if they would have gone in a different direction.

Honesty and perception management. They are often different things, and skilled corporations use the distinction to advantage.

7.3 Marketing and sales and the use of force

Marketing is all about creating perceptions in the mind of customers and potential customers that will tend to bring them to perceive you in the desired light. Sales on the other hand is about generating an exchange of something of value to you for something of value to them.

In most cases, the goal is to bring customers to buy things, and it is often considered better if they buy because they want to and like to rather than because they have to, but coercive force does, of course, work.

An example of force is the requirement to get a lawyer to represent you in court. While you don't strictly have to do this, failure to engage a good lawyer likely means that you lose the case. In which case simply not attending is probably better for you. The old saying applies: It is better to remain silent and be thought a fool than to open your mouth and remove all doubt.

Marketing can force the hands of others. For example, by creating the perception that you are the best and they are the worst, you almost force them to deny that they are bad, resulting in the appearance of defensiveness. When Kerry failed to do this in the 2004 election, he lost. Of course he also decided not to strike back with a similar perception campaign. In business, the same things can happen, but a more powerful form of force comes in intellectual property rights and the use of the legal system.

Patents are an excellent example of a big business strategy to use force against other businesses to create barriers to entry and to cause them to do a worse job of entry into a particular market. It is also used by lawyers increasingly to extract funds from enterprises that accidentally use intellectual property that was originated elsewhere. The legal system is one of the fundamental institutions of force between businesses. Civil suits are problematic and create large-scale public relations problems even for innocent parties.

7.4 Legal constraints and buying the law

So the legal system is the path to resolving disputes, but this system is fatally flawed when it comes to large businesses because it is unable to associate punishments with criminals rather than their businesses and because large businesses seem able to work the system through the handles of power and spending money to delay, limit damage, and otherwise slow the legal process to reduce or eliminate its effectiveness. When Bill Gates pays $800,000 in fines it is hardly noticeable to him. And when Microsoft is found guilty by the US courts in an antitrust suit the result is no punishment whatsoever. This cozy relationship between criminals in the corporate sector and government officials is problematic and eats away at the public trust. It also means that the dominance of Microsoft will remain because, even if you win against them in court, you still lose in the end.

And of course it is not only Microsoft, they are just the most obvious and high profile example. The scandals associated with Jack Abramoff demonstrate all to clearly how influence is peddled by enterprises of all sorts. It becomes extortion of a sort when you cannot succeed in a business without essentially bribing public officials or their friends by paying someone for access to the decision makers and a seat at the table.

The public perception of fairness in government contracts is all but gone in the US, and corporate influence is all too obvious. Fund a congressman for reelection and expect access to meetings, votes in your favor, and laws written to exempt you alone from specific taxes in specific years. Fund them some more and they will subsidize your business over your competitors. Do so as an industry and the government will support your whole industry. Companies understand pork barrel politics and peddling of influence and do so because the system forces it. And if this sounds US-centric, it is. Because in much of the rest of the world the same is true, but it is not the same hypocrisy. Try to get a contract in India or Russia without a fixer. It's all but impossible.

7.5 Internationalization and skirting local laws

Another favorite business strategy is the use of internationalization to avoid taxes and otherwise skirt laws. It's not a hard matter to understand that corporations don't have actual headquarters in the places where they are legally registered to avoid taxes. If the CEO of the company had to live and work where the corporation was headquartered, this would have a serious impact on companies all over the world who avoid taxes in this way. But the law does not follow such paths.

> It is essentially impossible to compete in business today unless you learn to game the system.

The use of virtualization is another form of information warfare. The virtual presence here or there for one purpose or another allows companies to avoid laws of all sorts, but lawmakers are trying to catch up. They are increasingly trying to extend laws internationally so that an act defined as a crime in one jurisdiction committed by a citizen of that jurisdiction elsewhere remains a crime. This is, of course, counter to all of the history of law. It means, for example, that if a citizen of the US goes to India and pays the customary fee for gaining access to a market, they will be violating US law. Of course this also makes it advantageous to use non-US citizens for these tasks, increasing the use of outsourcing and offshoring. The difference between a legal and illegal bribe in a foreign land is the citizenship of the employee.

Gaming is a fundamental part of business. Whatever the rules are, the successful businessperson finds ways to get around the ones that stop them from doing what they want to do. This is expected. But when the gaming extends to law breaking it creates an unfair playing field. After all, I simply cannot compete in India if I cannot pay for services to facilitate permits. So my competitor who does this wins all of the work and succeeds while I fail. I can choose not to go there but then I lose the advantage of low cost high quality Indian programmers, telephone answering services, and so forth. I don't mean to pick on India here, they are only an example.

Offshore gambling is an example of skirting the law. All you have to do is open a company somewhere that gambling is legal and have them offer gambling into the US. Those that participate may be breaking US law, but you are not and your company can do business with the customers willing to break the law or ignorant of the law. The same is more or less true of almost any criminal activity you want to name. It may be illegal here but it is often legal somewhere else. And with the Internet as the means of facilitating the operations, business can operate efficiently from anywhere in the world.

> Internationalization of business is one of the most dangerous things faced in information warfare by nation states.

Offshore data workers and privacy laws are another area where the law has not caught up with the realities of the Internet. As an example, when a state government wants to reduce costs for information operations, they might decide to outsource the data processing to a US company that in turn decides to offshore data entry and tax processing operations to China. China, of course, doesn't give a hoot about US or state privacy laws, and they will use the information for part of their ongoing information warfare efforts. They will identify and target US citizens of Chinese descent in financial trouble and working for a key company and appeal to their blood line and pocket book to get them to work for China. It benefits everyone, except of course it hurts us all. The offshoring is perfectly legal today, and it demonstrates just how easy it is to lose in information warfare.

Offshore and outsourced workers and the lack of reciprocal agreement requirements or the ability to enforce intellectual property or other requirements can destroy a company just as it can be a boon to efficiency and cost reduction. Remote workers lead to loss of control over content, the life blood of most enterprises today. Most enterprise executives don't realize that this loss of control is potentially devastating, but give them time. As major global enterprises start to fail, they will figure it out.

7.6 How will niches survive?

As big business exploits the advantages of information technology to dominate increasing parts of the business space and uses its large-scale buying power to make price competition impossible to overcome, there is a tendency for large enterprises to take over entire fields of endeavor, knocking out the small time operator and medium and small business until they no longer have a niche to survive in. At least this is the fear we see every day as WalMart and similar huge companies displace the multitude of less efficient businesses that have the misfortune of selling into the same market as they sell into.

> Information technology is the critical component that allows scalable efficiency for large businesses.

This theory has two interesting parts. One is that price competition will defeat all comers, and to the extent that service and quality are not significantly worse, it will indeed win out. In fact large volume businesses tend to have outstanding service and quality in their products and give great value for the price, and the market recognizes it as more and more people buy from their outlets. But the other half of the equation is the part I am concerned about here. That is the use of information technology to make it all work.

Efficiency has never been the hallmark of big business. As size grows, more levels of management are needed to handle the people and coordination becomes slower and more difficult to do. Internal quality control becomes harder as more and more people are in place and have larger and larger variations in outlook and behaviors. There will always be the potential for bad folks, up to the level of criminals, in any enterprise. As size increases, the likelihood of criminals and other nefarious characters increases in proportion. But information technology and sound business practices can make this quality control process even more efficient and effective than small business if properly managed and if information technology is properly leveraged.

Costs of background checks go down with volume allowing every employee to be more thoroughly checked out. Human resources systems provide increased vigilance and the integration of operations with information technology allows closer and more accurate tracking of employee performance. Fraud detection becomes more accurate and effective when full time experts can be employed, something that can only be afforded at a certain scale. So the question remains, of how the smaller company or niche player can survive in this environment and whether this situation will ultimately destroy innovation by making all niches too inefficient to survive. I think that while consolidation will leave small numbers of large players in some arenas, there will always be entrepreneurial niches

These niches are most likely to survive in areas where human expertise and ingenuity are key and where the corporate mentality will not succeed. The information protection arena is just such an area for some very important reasons. One of those reasons is that in order to be effective at information protection, you have to be willing to stand in direct opposition to top executives who are trying to cover up their misdeeds. If corruption is to be fought off, it can only be done with truly independent reviews, such as those that are supposed to be provided by internal audit functions. But I am probably blinded by my own profession.

In the information warfare arena, nice advantages have always been key to success. Specialized deep expertise in one area can often defeat large-scale mass and large numbers of forces in another arena. Battlefield deception creates asymmetric advantage just as it does in niche businesses which can push an area that the large enterprise cannot or will not push forward. This is done at greater risk to the entrepreneur than the enterprise is willing to accept or in an area the enterprise has not yet considered worthy of investment. Or perhaps it just hasn't gotten properly prioritized by the large entity. So the niche player will survive by taking higher risks in narrow areas and build themselves a survival strategy as they always have.

7.7 Corporations against each other

Corporations are always battling each other in the information arena. Sometimes these battles take on a public face, but in other cases they are private battles and even extend into direct malicious acts against each other. As the intensity of these conflicts increase, eventually we reach a level of intensity generally associated with information warfare if in a different context.

> Corporate intelligence can be readily taught and carried out legally or illegally.

Among the ongoing battles between enterprises are the corporate intelligence gathering operations that are commonly used. There are two general types of efforts undertaken in this arena. Open source intelligence is used to seek information on enterprises using what is openly published. This is basic work that every company should be engaged in against competitors. It provides the context for decision-making and is used from training sales staff to strategic decisions. I have engaged in this work for enterprises wishing to understand markets or make strategic decisions about entering markets or acquiring companies. This is reasonably fair and completely legal.

The other type of intelligence gathering is surreptitious and involves deceptions of various sorts, break-ins to computers, and corporate espionage tactics. I have done this sort of effort as well, but only for corporations against themselves. Otherwise it would be illegal in most countries.

The intelligence process can be relatively easily understood at a theoretical level, and with training and effort, we have found that students can become quite adept at these sorts of operations in a relatively short time frame. A semester or two of graduate education plus some real world experience and a life time bent for this sort of thing helps. The basic attack and defense tactics and methods are taught as part of the graduate deception course I teach and can be understood by reading *Frauds, Spies, and Lies*.

The basic process we use starts with a matrix of information we are trying to get and a set of techniques we are willing to apply to get the information. From my Web site we even sell assessments that evaluate susceptibility to:

- **Dumpster diving:** We go into your waste baskets and garbage areas and collect and analyze what we can glean. We often get lots of corporate secrets and personal identifiable information.
- **Web-based Intelligence:** We download everything we can find on you from the Web including not only your sites but all other Internet information. We commonly find every location of the business, lots of employee information, what kinds of systems are used for what, executive salaries, and so forth.
- **Deception Susceptibility:** We use deceptions to gain entry, access, information, and so forth across a full spectrum of issues. We are often able to rapidly gain a lot of internal access and confidential information.
- **Telecommunications Sweep:** We sweep telephone, cellular, fax, modem, and Internet components of corporate information assets. We often finds scores of weaknesses that can be exploited.
- **Project Data Aggregation:** We try to gather information on a particular project to find out what should be secret. We often find out a lot about projects that the enterprise thought were very secret.
- **A Bugging Exercise:** We listen into conversations in executive suites or other similar places by planting surveillance devices. We usually succeed in gaining a lot of high-level information.

These are studies we perform because corporate attackers perform them against enterprises to gain advantage. Ask Oracle about it if you don't believe me. They were caught dumpster diving against competitors. Of course the idea of these studies is to validate the vulnerability and help figure out what to do about it.

7.8 Corporations against nation states

Corporations today often find themselves engaged in information warfare of a sort against nation states. While this may seem to be asymmetric, nation states are catching up and have a chance if they act quickly and think through the process.

Yes, that's right. Large multinational enterprises have serious advantages over nation states in information warfare. They have assets distributed all over the world, a lot of diverse expertise, they can act quickly and without much political fallout, and they can engage other nation states in the battle on their behalf.

> The corporate information war with China pits capitalism against communism in a fight for the hearts and minds of the Chinese people and information dominance for China over Western technology infrastructure.

These battles are underway and have been for a long time. China is currently being most aggressive in this area and is succeeding where France seems to have failed in their attempts over the years. They are leveraging their low cost labor and large market to force enterprises into business relationships that grant China access to internal secrets and technology bases and position them to take what they are not granted.

But the battles are hardly one sided. Companies have long exploited their presence in countries to take intellectual capital and influenced elections and policies around the world. In some sense the larger battle going on in China is representative. While China seems to be winning the access to information, the Western societies are also trying hard to push national philosophy and capitalism to replace communism. And to the extent that material goods are increasingly desired and acquired in China, the corporations are winning that part of the war. The hearts and minds war is an information war that is being carried out as national policy by the US and many other nations against the Chinese communist government and it is as pure a form of information warfare as you can find.

World War 3: We're losing it...

Capitalism and corporations have long been at information war with the nation states of the world. And while the outcome is far from decided, it seems clear that the corporations have some major advantages in today's environment that, if pushed, could result in the long term corporatization of the global political system in a fashion similar to the future depicted in the movie *Rollerball*. Corporate dominance will likely not be oriented toward division into energy, food, and so forth, but rather there will be surviving niches with a small number of large competitors and lots of niche players.

> The information war of our age may be the war between capitalism and all other forms of government.

Capitalism vs communism is not the only battle underway. The battles in the Middle East are also corporate wars that pit religion against corporations. Will companies be able to corrupt the remaining religious leaders by making them rich? It has worked for many years in many of the countries in the Middle East, except that the mix of dictatorships, even when family run, seems to fail against a republic when it comes to maintaining control over the people while rewarding those who support the corporate approach. Money, the easy life, health, survival for your family, power, and all of the trappings that come with it are the rewards that corporations can bring to those willing to use power in their favor. And this also benefits the people who live in those countries whose standard of living tends to increase along the way.

But overcoming religion and indoctrinations over generations is a hard thing to do and it will take all of the assets and resources of large enterprises and the governments that support them to overcome these barriers and win the information war. Do any of us want the world ruled by a few uncontrolled dictators who happen to wear corporate shirts instead of robes? Or can we form a global future that includes provisions for controls over monopolies and keeps competition and progress alive while preventing outright exploitation of people? Only time will tell.

7.9 The military information complex

The military industrial complex that Eisenhower warned the US about has largely come to pass, and the hawks are currently dominating in the US. But this is likely a cyclical thing that will wax and wane until the participants realize that it is the information war they should really be fighting and winning. The hearts and minds war, once it becomes the focus of this complex, will increasingly dominate the battle space, reducing the human suffering for those that have advanced weapons. While the small unit operations are slowly eaten away at, any hope of large-scale warfare is already nearly dead without information technology highly supportive of the effort and the industrial base present to support it.

What can the complex do for the world and how do we have to control it?

This then is the battleground for the corporate war with religion and communism and the other isms of the world. Will the corporations of the world become inundated with religious fanatics who end up ruling the roost, or will they inundate the world with increased well being and better lifestyle? If and when the corporations take over the world and grab control over the still unfettered areas of population, will they be able to maintain control for long enough to transform the societies over generations and eliminate the religious fanatics altogether? These are the deeper questions of corporate information warfare that need to be answered and upon which hinge the future of the world.

The path to this end lies not in the military industrial complex, but in the military information complex that is not yet formed. It is the complex that will win or lose the hearts and minds war and it will involve military and industrial elements from all over the world. It is one of the key elements that will support World War 3, assuming that the world can go down this path. But this is not such a simple matter, and the shape of the path and the resulting global society is in deep question. Unfettered power in competitive corporations is not exactly a desirable future.

The future of the world and its people depends more than ever on the future of corporations and less than ever on the future of nations. Unless and until someone is insane enough to launch real weapons of mass destruction in large volume, the world will move down this course of information warfare in increasing increments with corporations ruling the roost more and more over time. Unless the nation states decide to defend themselves, the corporations will take over, but how can the nation states defend themselves? They depend on the corporations for increasingly large portions of their capacity to exist and to rule. As one famous economist once told me and a room full of people, if nation states didn't exist, corporations would invent them to deal with things they don't want to have to deal with.

> Government by, for, and of the corporations shall not perish from this Earth.

But this point of view runs opposite to all of the history of humanity, or at least partially so, and the question must arise of whether it can in fact be realized. When the US was founded, and when most of Europe arose from World War 2, there was a global sentiment around the good role of government in carrying out the desirable elements that it brought to the people. But the view of many in the corporate world is reflected in the corporation rather than the people as the objective of government. The corporations are the fifth estate of government. Legislative, Executive, Judicial branches of government, the press, and the corporations.

But the corporations are moving in on government, and the US is the perfect example of it. They increasingly control the legislatures either directly or indirectly by the funding and influence they use to get their folks elected and threats of removal from power. The executive branch in the US today is already occupied almost entirely by big oil executives. And the courts can be controlled by a multi year strategy of controlling the legislature and executive. Only the press remains, and it is increasingly controlled by large corporations as well. Where will it all end?

8 Propaganda

I covered propaganda at a relatively light level in *Frauds, Spies, and Lies*, but I hope to cover it in more depth here. I will start with an extract of that book, suitably modified, and without further citation.

Propaganda is perhaps the largest scale and most harmful sort of deception used on populations. Webster's dictionary says:

> *"(1) The spreading of ideas, information, or rumor for the purpose of helping or injuring an institution, a cause, or a person ... (2) Ideas, facts, or allegations spread deliberately to further one's cause or to damage an opposing cause; also : a public action having such an effect"*

> Hitler's propaganda machine not only won the people and helped him start World War 2, it caused the death of millions, got ordinary people to kill each other without just cause, brought misery to the world, and it still lives on, even today, long after the death of those who created it.

Hitler's Germany built up one of the most powerful and horrific propaganda machines in modern times. It created levels of hatred never before achieved. It started by declaring Germany oppressed, moved on to seeking common enemies in the Jews and Gypsies, who were proclaimed to be supporting and controlling governments all around the world (sound familiar?), and ultimately sought the systematic murder of millions, euphemistically called the final solution. And even when defeat became inevitable, they wanted everyone to stay the course, declaring that they were winning all along, even as Hitler eventually killed himself. How long did it last? It lasts still, as the Hitler propaganda machine generates denials of the Holocaust, Hitler's burnt body was never confirmed as his by any adequate scientific means, and Nazi groups around the world still persist in claiming that the Jews control the world governments.

Of course propaganda examples are pretty easy to find in any society, but they rarely go to the extremes of World War 2, except when conflict levels increase in intensity. But how do they increase in intensity? Through the propaganda machines that fuel them. Propaganda is the state's main direct instrument of information warfare.

Brian Standing recently released his video titled "*War i$ $ell*" that outlines pretty well the elements of a propaganda plan:

- Demonize the enemy
- Get 3rd party endorsements
- Use branding
- Stay on message
- Tell the Big Lie
- Use doublespeak
- Silence the opposition

Another way to look at this is by following the typical marketing strategy; create a need and fulfill it. In other words, in order to get people to do what you want, create images in their mind that cause them to believe that it is important for them to do something you want done, then provide the path to getting it done.

- **Hitler**: Create the perception that the economic crisis is due to a common enemy that can be found at home and abroad and that can be clearly identified by physical characteristics and strange practices; the Jews and Gypsies. Then induce actions against them to achieve your goals of power; Crystalnacht.
- **Bush**: Create the perception that the attacks on the World Trade Center were due to your enemy; Iraq, the objective of a desired war from the time your team got into office; and augment it by the threat of imminent massive attack. Then induce actions against them; start a war.

I am not trying to compare Bush to Hitler, just to put things of today in historical perspective. But some in Iraq are.

8.1 Identify and demonize the enemy

Identifying an enemy is fundamental to creating a war. Without a common enemy, it is hard to solidify the public into support. Any enemy will do, but it is a lot easier if there are identifiable differences between them and us so we can use those differences to make it alright to hate them. And make no mistake about it, the goal of propaganda is almost always to create hatred and take advantage of it. The only real exception is the cult-based creation of the charismatic leader, but even this typically includes hatred for someone else along the line. Without conflict of some sort there can be no urgency for action. Think about how to make a good movie – it's the same thing. You need conflict and it has to be resolved by the action in the movie. Plot is the key and theme brings the emotional content.

> Propaganda needs an enemy to get the populace to act.

A great example is the story about the first Gulf war when the Iraqis were accused of removing babies from incubators and letting them die on hospital floors. It was a lie according to the doctors in the hospitals. But it was repeated by US senators, President Bush (1), and others, and was a rallying point for demonizing the Iraqis and increasing the urgency of the war.

Of course the counter to this approach is to humanize the enemy. *Those poor Iraqis that are just trying to earn a living and improve the lives of their families depend on Saddam Husein to keep the evil hordes of Iran away.* In this process I have also created a new enemy, Iran, and after all, the enemy of my enemy is my friend, isn't it?

The need for differentiable properties is fundamental. Without something I can look at, hear, feel, smell, or some other visceral dimension to it, it is hard to create hatred or even tell who the enemy is. The only thing to fear is fear itself is the way to back down from this, but mostly propagandists want to create fear by creating and giving super powers to enemies. There are a lot of standard techniques to do this.

In most wars a name is given to the enemy that makes them seem less human. "Gooks" from the War in Viet Nam, "Huns" from World War 1, and so on. It's a lot easier to kill people when you don't think of them as people. They are just demons.

Race is a good way to demonize people because it is so apparent from looking at them. Accent is usable but it is harder to push through than race because there are so many variations. Smell is a good one too because food, which is cultural, also produces smell, which allows a clash of cultures to be cultured. Germany's Jews and Gypsies were easily recognizable next to the master race. Arabs are easy for the US to recognize, except of course that the US has a very mixed population.

> The propagandist is a divider, using divide and conquer as a technique except in the rarest of cases.

Divide and conquer is the watchword of the propagandist. And while it is getting harder to use racial lines if you are from a Western society, it will still work for much of the world. Demons have exaggerated characteristics in posters and similar material that the propagandist uses. Their noses are bigger, eyes closer together, and clothing bears dark marks and clear coloring. They have the evil look about them.

Words are associated with demons as well. The axis of evil associates World War 2 Axis powers with Iraq, Iran, and North Korea and implies they are a coalition even though Iraq was worst enemies with Iran for a long time and North Korea has nothing at all to do with the Middle Eastern countries. The association of Islam with terrorists is a good example of a religious association with terror. While it is true that some of the terrorists are radical and claim to be Islamic, this is really part of a class struggle in which radical religious views are used as the only workable banner under which the radicals can employ their propaganda. In that region of the world, class struggle won't play well and religious leaders are required in order to motivate actions such as suicide.

8.2 Gaining endorsements

Propaganda from a single source tends to fail. In order to be effective, it is necessary to create at least the image of a consensus. Therefore, a consensus is created in any way available. In Iraq today, the US is getting endorsements by buying reporters off and sending them scripts on what to report in the media. This sort of thing fails if poorly done and the US has clearly demonstrated their lack of prowess in this arena with the foreign press. Endorsements that are less poorly done obviously work a lot better.

More effective endorsements are viewed by the target audiences as being independent, knowledgeable, sincere, and diverse in their backgrounds across the space of the targets. Using the basic principles of influence as described in *Frauds, Spies, and Lies*, it is clear that for each audience, the better endorsements come from someone that looks, talks, and acts like them. So with adequate targeting, endorsements can come into every home from the proper sources.

- **Independence:** The press and other media is widely viewed as independent in certain areas of the world, but as they are abused more and more, this is going away. A professor from a university in the proper area of expertise is often viewed as independent and knowledgeable. But business people are not viewed as independent because they do things for money.
- **Fame as an endorsement:** Famous people tend to bring credibility, unless they are famous criminals, and even then they seem to be widely brought in as pundits in their area of crime. Endorsements from actors and activists and those who have been widely quoted in the media bear more weight than others.
- **Athletic endorsements:** Race car drivers work in much of the US South and in certain areas of Western Europe. And nothing compares to a football star joining the armed forces and going to war, unless they die from friendly fire.

- **Sincerity is worth a lot:** That's why a mother whose son died in a war is played up to great effect in a war protest. She is perceived as having a right to complain even if others do not, and people will excuse what might be considered impolite behaviors in others from someone who has lost a relative, but only for a period of time.
- **Religious leaders:** These leaders who have almost nothing relevant to bring to bear in many areas are given enormous credibility because they have the perception of honesty, sincerity, and caring attitudes that most of us can only admire. This is one of the reasons that the separation of church and state is so critical to politics, and it is a reason that religious leaders that get too embroiled in politics end up being marginalized.
- **Funding:** The saying is that money talks. And it most certainly does. Money can buy plenty of endorsements and plenty of credibility if properly applied to public relations.
- **Introductions:** If introduced as an expert, you will be believed as one. Of course you can blow this by removing all doubt, as the saying goes, but for the most part, you will be respected based on how you are presented. A proper introduction followed by a proper presentation becomes a very powerful endorsement for a point of view.
- **Associations:** Associations such as name dropping often lead to believability even though it makes no rational sense. After all, who are you to tell me that what I told the President is not good enough advice for your group?
- **Peer groups:** One of the best ways to get a viewpoint across is to present it as coming from a peer group. This is especially powerful in the Internet where peers can be generated for groups with a minimum of effort and with little or no validity behind them.

Endorsements of all sorts can be generated with relatively little effort and they are powerful methods for generating increased credibility and liking for your propaganda campaign.

8.3 Branding and associations

The idea of branding is to stamp sayings in the mind of the targets. For example, linking support for a war to freedom and democracy even if the war is not about that at all, is a form of branding. By branding, an association is created between things.

Associations are powerful mental and psychological mechanisms. Dogs can be made to salivate at the sound of a bell by association and people can be made to think of things they like or hate by association. Association is a fundamental of behavioral control and it is used to manipulate and control the behavior of people, animals, and organizations.

Most propagandists use existing associations and extend them to create new ones along the way. This is far easier than trying to create associations from scratch, although over long time frames, this can be accomplished reasonably well. In creating these associations, the typical process starts with something yielding similar emotional responses to the desired ones.

- If you want to stir up hatred, use something hated and associate it with the new item. Like associating the Soviet Union with the *Star Wars* evil empire and associating the well known Pink Floyd album and movie *The Wall* with the Berlin wall.
- If you want people to love their charismatic leader, associate the leader with flowers and spring and love and good food and pure water.
- If you want people to rally round the flag, associate the enemy with flag burning and desecration of graves of soldiers draped with your flag.

Think first of the desired emotion, then create an association with words, sounds, images, and other senses between the desired state for the desired target and the item you are trying to portray. You get propaganda, or advertising, depending on your objective.

The portrayal of the political system as a sort of religion is part of the association some politicians try to use. Religion is a deeply impressed part of most peoples' lives. Let's try an example of a religious view and a desired state; the Hindu belief in reincarnation and their reverence for cows to be associated with hatred of the Japanese (I have no particular reason to choose this pair, it's just an example). You portray it, perhaps, as a Japanese person with exaggerated and highly violent appearance but otherwise racially Japanese appearance in the face, with blood coming from their mouth as they emerge from the grave with their bodies resurrected as a giant and horrific bug that is ripping a cow to shreds. The Japanese associated with cow killing blood bearing horrific bugs will most certainly be an association that the Hindus will revile.

That is the imagery of propaganda. Now you add a saying that relates this Japanese beetle with pestilence of the sort that has caused the most recent famine in India. Blame the Japanese biological warfare program (Is there one? There was in World War 2, so we will resurrect it in our propaganda!) for the rampant disease in some part of India that is killing many. Put up a voice of a poor dying Indian child crying out to its mother and have an interview describing how this young child, like so many others, has died of a mysterious disease now spreading where the Japanese consul recently visited. The rumor mill is ready to explode and the conspiracy theorists are chomping at the bit, so we will encourage them with a bit of Internet posting to select bulletin boards and perhaps an interview on radio with someone from the government that is denying that there is any proof of the link between the well known Japanese biological warfare program and the deaths of Indian children and families where the Japanese ambassador visited. A well planted news question asking about the rumors of new racially targeted weaponized diseases being developed in Japan at (name a Japanese institute for medical research) and you will have a full fledged propaganda campaign made to order.

> A well planned set of messages applied in the proper context and sequence can create a lot of force.

8.4 Consistency and messaging

I particularly want you to note that in the last section I didn't tell a single lie, nor did anyone who I put up to speak for my efforts. Look closely. There was a Japanese biological weapons program in World War 2 and it was used against the Chinese. Nowhere did I say that they had an ongoing one. The attempt to blame the Japanese need not involve lies, only rumors, which the government denies with complex and incomplete denials, thus legitimatizing the rumor. Add in the planted media questions and the posters which don't even try to state any facts, and there are no lies at all. Conspiracy theories about how what is now assumed to have happened could have happened only fuel the fire without adding any new facts. But now the need for some set of consistency becomes apparent. If the messaging gets too inconsistent, it will end up confusing and become untrusted and decried.

> The script allows the players to be coordinated toward an end and the message to be clear and consistent.

What we need is a good script for this play. If you haven't seen *Wag the Dog*, it's worth getting the movie and watching for an outstanding and outrageous example. The need for a clear script is fundamental to getting desired emotional responses from the audience, but unlike a two hour presentation at a local theater, propaganda is a play for the masses and has to last for months to years in order to have the desired effect. The script is not all written at the start, it has to be adapted to the circumstance. So you need to have a set of script writers that constantly rework the themes, create new images and phrasing, and adapt to whatever takes place.

Scripting can be done well and subtly, or poorly and obviously. As scripts become too standardized, they become very obvious propaganda, and even with media cooperation, the people will start to see through them. Standard answers to questions where the same phrases and changes of phrases are used by an entire team of politicians are clearly scripted and bring about disdain.

The skilled script follower will not simply take any question that is not aligned to a standard answer and twist it in the response toward a standard answer as so many are trained to do by their public relations specialists. Key phrases may be reused but not repeated time after time after time. While this works for some amount of time and some amount of reuse, it can be overdone. The key is to make it look natural, and the problem is that this takes smart people who are willing to take the time and effort to focus on the propaganda and invest their time and effort in its most adaptive form. These folks are called press secretaries.

There is no doubt that consistency and repetition drive messages home. They are effective at generating memories that are clear and to the point. They create patterns in peoples' minds that are like the song you can't get out of your head. But they can also cause people to stop listening and come to oppose you. The Bush (2) administration in the US is the all time best example of this and the media responses are classic.

Bush went a bridge too far and got nailed.

Bush got the benefit of the doubt for years after the September 11, 2001 attacks on the US, and the Iraqi war was treated as an extension of it. Phrase after phrase was repeated by everyone at high levels in the Bush administration in seeming lock step and the media played it all up and almost never brought it out until after the second Bush election. But as soon as the second election was over and the money stopped flowing to the media outlets from the election campaigns, the love affair ended. Within days of the election ending, truth started to emit from the media in dribbles. Then, when Hurricane Katrina hit and thousands of people died from government inaction, the media was right there, and they felt betrayed. From that moment on, it became a slug fest. The media no longer accepted anything they were told, but started to challenge things. Repeated phrases were identified every time the words changed and one quote after another was played next to one another to show how the lock step process worked. Bush's popularity and credibility plummeted.

8.5 Combine fear with action for effect

For best effect, propaganda must be adaptive and seem to be natural even when it is not. The naiveté of the audience must be exploited and the presentation must be well done in order to move the masses slowly without breaking them up along the way and creating fractured support. With a substantial opposition party available to shoot things down and the media willing to play them almost as much as you, how do you do this? The answer is that you need to combine extremes of fear with actions that can be taken to mitigate the fear.

Psychological studies show that the more intense and immediate the fear, the more effective the message is, so subtlety is not a good approach in terms of telling the population they will die a horrible death at the hands of the evil ones. Just come out and say it using the demonized naming and messaging described earlier. But for high effect, action must be available.

> Psychological studies show that intense fear combined with available mitigating action produces desired behaviors.

When people get really scared, they have a tendency to stay at home, stay in bed, and not want to venture out. The fear after September 11, 2001 in the US generated so much of this that the media and the leadership had to tell people to go out and travel to save the decimated transportation industry. The devastation came from fear without direction. In England after the 2005 subway bombings, there was a day or so of slowed transportation performance, but the government acted quickly to show things were safe (regardless of realities) and the economic impact was minimized. In Spain, after the train bombing a few years earlier, the election was swayed and the candidate that the terrorists wanted to win ended up winning when they otherwise would almost certainly have lost.

Psychological studies over decades found that the availability of action to mitigate fear leads to that action being taken.

So the propagandist who really knows what they are doing will not just send a message of fear. They will augment fear with desired mitigating action to induce desired behaviors in the populace. Returning to advertising for a moment, we see the clear connection. Advertisements seek to create a need then fulfill it. Propaganda seeks to create a need and then fulfill it. Advertising executives have long understood that sex and fear of death sell best. After that come being better than the Joneses, the grass is always greener, and life crises. Everything else is almost not worth the effort.

> Propaganda is very much like dog training. Encourage a natural behavior with positive feedback in the form of a reward and then associate it with a command so you can call on it when you want it.

The ideal propaganda might have iconic male and female stars of stage and screen inducing fear of death and solving it with sex. Just consider how easy the sale would be if the threat was biological warfare and the solution was a dramatic increase in population and population diversity. A government sponsored rush to have sex with anyone you could find is the only solution to the advancing biological weapons program of the evil empire! Sadly the religious groups will never go along with this one, unless you can make it God's will by finding the right scripture to twist and a few religious leaders that back it up with newly created official doctrine.

That is the nature of propaganda. In essence, you can cause people to do things that they would never consider doing without the propaganda and it is all the easier to bend their will toward what they would naturally do in some circumstance anyway. Find a natural behavior and control it by supporting it through positive feedback and associating it with a command. If this sounds like dog training, it is. This is exactly how you train dogs, and people too. In propaganda these same principles are applied using psychological results to great effect in order to influence large groups to behave as desired.

8.6 Sustaining the big lie

The big lie is the lie told by people in a position of authority as if it were fact when it is in fact a lie. If they repeat the lie enough times, are in authority, are trusted, and go relatively unopposed, the lie will become a social truth and be treated as fact. If this seems like a rough way to go, it is, but if you want to do something like start a war without a reasonably just cause and against the longstanding historic norms of your people, there is little choice but to use the big lie. And of course the big lie is really and truly a lie, not just a belief that is hard to shake. The big lie is something that the originators know to be a lie but stick to anyway.

> There's no accounting for taste, but taste seems to be acquired rather than built-in. Rather than try to change the flavor, you can change the taste of those who eat it

Advertising has its brand of the big lie, and it is one of the increasingly common things we see in advertisements. When there is a negative associated with a particular part of a product, rather than try to shy away from it, the company will advertise their way out of it by calling it a feature, advertising a standard feature as if it were a special feature, and trying to associate it with safety or sexy things to create a desire for it rather than a disdain of it. The exploding gas tank of the Ford Pinto is countered by good mileage and price claims and depicting it as the ideal car for the young family. The Jeep's tendency to roll over is countered by showing it run rapidly over rugged terrain carefully designed to not cause it to roll over while looking like it should. If your product doesn't meet the current standard of beauty, change the standard!

How many times did we hear that there was no question the Iraqis had Weapons of Mass Destruction? The more this is investigated, the clearer it becomes that the evidence was all against this position, but *don't confuse me with the facts* became the watchword. Who tells the big lie? Whoever is in power! Nobody else can!! They would if they could.

8.7 Doublespeak and other speech patterns

Phrases like the "axis of evil" are outstanding examples of wording that implies both the evil group of enemies in World War 2 and the notion that the countries listed are somehow working together even if they are not. But doublespeak is only one of the many speech patterns that are used by propagandists. As the literature in psychology has long revealed, using the same terms as the audience causes them to like you and believe you. But the propagandist has a problem in a diverse society in that there are so many cultures involved that favoring one may offend others.

> Keeping it simple and basic is best for national campaigns. The cowboy thing works in the US and the majestic works in the EU.

Common themes are better for propaganda. Technology makes it increasingly possible to tailor messages to audiences. But while elections now do custom mailers for individual household members by targeting just the issues they are interested in according to the intelligence process applied to their medical, cultural, and purchasing history, this sort of funding and attention is not available for normal government operations and would be considered inappropriate by watchdog groups. So a far broader brush stroke has to be used for national propaganda campaigns. Thus the simple basic – fear of death.

While saying don't torpedo this project to an admiral and don't shoot this project down to a pilot might be useful in special contexts, the same does not work across the full spectrum of the US or the EU. China, Japan, India, Pakistan, and lots of other historically ethnic countries have a far easier propaganda environment to work in than the US and the EU. In the US some social commonalities exist for coalitions of people, which is one of the reasons that the cowboy approach can function in politics. Being a cowboy has girl appeal, southern appeal, regular guy appeal, and movie associations. But it doesn't work well for the liberals in the country and creates opposition. So the counter to the opposition comes in the form of marginalization.

8.8 Silencing the opposition

A key factor in success of any propaganda campaign is reducing the audience resistance. This is done in many ways, and most of the issues addressed to here have been associated with making messages more appealing and easier to apply. But in most such campaigns, there is some level of opposition. In order to be effective, the propaganda campaign must find a way to reduce the effect of opposing voices. Otherwise, the campaign will fail.

The classic example is that war propagandists attack anyone who opposes the war by challenging their patriotism, calling them names like pacifist, marginalizing their views as right or left wing (the wing being the key thing that sets them far off to the side of us centrists and radicalizes them), by saying that they are doing nothing while evil flourishes, and by creating an us and them thing where we are all us and only they are them.

> Like any successful religion, opposition must be silenced or rationality will win out over the dogma of propaganda.

The phrase "If you aren't with us, you are against us" is very similar to "The enemy of my enemy is my friend" except that the former is Western culture and the latter is Arabic culture. They are both intended to quell certain contradictory views. The notion of the loyal opposition is lost, which often leads to catastrophic failures. If you are against the war, you are against democracy. Plug in your religion (ism), political system (cracy), or other "ism" or "cracy" and it works just as well. While the appearance of open debate must be maintained in order to keep things from looking too dogmatic in Western cultures, in most Eastern cultures, any vocalization of opposition is considered impolite in public. "*Never let anyone outside the family know what you are thinking*" says Marlon Brando as Vito Corleone, the head of the Family in *The Godfather*.

There are a lot of workable techniques for silencing or suppressing the opposition in propaganda campaigns because of the broad powers of governments in times of crisis and during wars.

Some of the best and most skillfully applied techniques are listed here as examples of how to silence opposition to propaganda campaigns:

- **Don't let them come to the table:** By having meetings on subjects of interest and not bringing them to the table, you keep their views out while building a defensible expert consensus.
- **Don't fund their science, fund yours:** The power of the purse strings works wonders. You can even get scientists to delay publications by threatening or promising funding, and of course by funding them in a different area you cut off their research in the area you want to suppress.
- **Use the chilling effect:** The story of what happened to the other guy who didn't go along to get along works well in chilling further work in any desired area of defocus. Think McCarthy-ism.
- **A few strategic arrests or tax reviews work wonders:** In suppressing opposing views, keeping them busy with legal cases or tax reviews always helps to distract them.
- **Make some folks disappear:** A simple disappearance or two has a really strong effect on reducing the vocalization level of the opposition.
- **Keep them off the television and speaker's tables:** Keeping the opposition out of the mass media and widely trusted forums is vital to success, and all the harder with the Internet providing so many pulpits.
- **Use positional power effectively:** Use your power of position with all of its might to keep information from getting out and to control the right to organize others.
- **Make up names for them:** Them hippie commie fags who hate America worked. Try your own! It's easy!!

Most importantly, do not quit, even while you are ahead. Don't give up the struggle against truth, justice, and morality, for if you do, the opposition will come and get you in the middle of the night like they did to my neighbor Jim... he never returned!!!

8.9 The end of news as we knew it

In the US, news was once something generated largely by reporters who sought it out and reported on it. It involved a non-profit element of the typical broadcast or print media that was supplemented by commercials but never really turned a profit and was not really intended to. It was the thing that drew the people to you because they wanted to know what was going on in the world. That time is gone now, a distant memory of the past.

News today is brought to you by those with the money to promote themselves into it. Reporters don't go out and get the news, they read it from news releases sent to them by those with the money to support the activity. If they are really go getters, they call up the people identified in the press releases and ask them questions over the phone while largely quoting the stories pre-written by the PR departments of the releasers.

And there is a whole industry that lives by this press release mentality. If you want to send out a press release to 40,000 or more media folks, Majon international will do it over the Internet for about $500. If you want them to write the release for you it will cost a bit more. These folks and their many competitors are professional, well thought out, and timely in their delivery of news to the media, and that means that the media need not think through things or dig deep to find stories. After all, once the 60 minutes on the news channel are consumed for the day, or the pages of the paper are filled, more news is just waste.

One of my current occupations is as an industry analyst. That means I get press passes and am considered like a member of the 4th estate by many in the industries I cover. So I get to see it from the other side as well. I get email after email offering to brief me on the newest, greatest, best, latest, only, only other, you name the superlative, product or service related to information protection. I can get pre-release notices of upcoming products and services before the rest of the world knows about it. And I talk to the CEO or CFO of some of the largest companies in the world.

When I go to conferences, I get a press pass. Typically that means I can go to any event anywhere in the venue at any time. It also means that I get to sit in the press lounge where there are meals, snacks, network access points, brochures of every sort in large numbers, and meeting rooms – lots and lots of meeting rooms. I meet with vendor after vendor, talking to their CEO, VP of marketing, top technical lead, or you name it. And I get all the news that's fit to print and a lot more.

The problem is that they are competing for the most precious resource I have, my time. And with the pressures in the media to deliver reports on a deadline, time is of the essence. The best reporters seek out second sources and ask other experts or analysts for advice or comments on the press releases they get from others. The really good ones dig a bit to see if these things are real. But one way or the other, they are not seeking out other stories, they are writing the story on the press release topic. They are researching but their research is focused along with their attention on the things that are delivered to them instead of the burning issues of the day.

> The press does not seek out stories. The stories seek out the press. So most of the most important stories never get told and we live in a veil of ignorance.

The Internet offers a unique venue for finding out more about more things in more places, but even this venue is not reflective of the world as it is unless those who contribute to it are of such a bent and have the time and energy and support to pursue it. In truth, the vision of the media as idealized by the media was never anything more than a pipe dream. And as the media slides down that slippery slope, it becomes the propaganda machine of the rich and the powerful. It supports their ideas, but more importantly, it steers the discussion, and as a consequence, gains control of the debate.

The media, in the end, controls what we think about more than what we think about it. And this is more dangerous than anything.

8.10 Recognizing and defeating propaganda

Of course in any war there is offense and defense. Defense against propaganda is pretty straight forward. Except of course it may be hard to accomplish.

- **Recognize the propaganda:** The first best step to recognizing propaganda is to assume that anything a government tells you about a war they are trying to promote is propaganda.
- **Demonize the demonizers:** Compare the propagandists to Hitler, at least in Western societies. In the Middle East, compare them to Jews or President Bush (2).
- **Humanize the demonized:** People are people, not sub-beings, not short catchy name things like *Glags* or *Slats.* Get them on television expressing centrist viewpoints and calling the propagandists propagandists. It's harder to sell niceness than fear, so...
- **Sell fear of propaganda:** Yes, that's right, you need to make people fear propaganda by pointing out how many people have died and how whatever they are selling you is not the biggest thing to be afraid of.
- **Attack their motives:** Try to associate profit motives with the propagandists – they are doing it for the oil revenues. This will work in some societies but not in others. Find the right analogy or rationale and push it hard.
- **Challenge their false facts:** Try calling them "false facts", rumor, hyperbole, ad hominim attacks, rumor mongers and phony experts. Get real experts and real facts, and do it fast. Challenge them at every turn and in real time and in public.
- **Endorsements are easy to get:** I got a letter telling me that I could get Alexander Haig, one time US Secretary of Defense, to endorse my company on a video that they would make for only $16,000. Rest assured that I don't know Alexander at all and he knows nothing of my products or services. That's how hard it is to get an endorsement.

- **Point out the fallacies and branding:** Branding is for cattle, not people. Don't be cowed! How's that for a saying. Use it against the cowboys and while you are at it, explain that the cowboy attitude is fine for chasing cows around but will cause pain, death, and financial crisis for the real world. Or try pointing to them as the past and you as the future.
- **Call them names:** You need good names that stick. And when they call you names, expose them as simpletons who are painting everyone in the country with a broad brush. You get the idea...
- **If they all say it, it is a script:** Call them script kiddies and point out every time they all say the same thing and every time they change their point of view as a group. Ask them embarrassing questions about the script and whether it was produced by the government propaganda office. When they avoid questions, ask the questions again and again. And start answering it for them. "I take it you agree then that you are a crook?"
- **The only thing to fear is fear itself:** This is one of my favorites. The implied endorsement of a past president.
- **Bring voice to the loyal opposition:** When the loyal opposition is being silenced or when everyone you talk to seems to agree, it's time to voice other views. You don't even have to believe it yourself, and you can portray it as what you heard or what you were thinking, but give voice to the opposition.

There are plenty of other specifics but the basics remain the same. By simply making a second view heard loud and clear and calling everything, you marginalize both sides, and this is fine for destroying the corruptive effects of propaganda. But be careful. If your goal is to bring a different side to the fore, the counter propaganda approach will marginalize you as well. It's best to outsource it as the propaganda folks in governments all over the world have learned to do. Then you can claim it wasn't you and that you are the truth teller to be listened to when the dust settles.

9 Politics

This of course brings us to politics. It has been said that politics is war without the appearance of violence. I don't know who said it, and I may have just made it up, but we are talking about politics, so anything goes.

At its heart, politics is and is supposed to be information warfare; war without the violence. While some political campaigns have lots of violence, most of them are far less violent than most military takeovers, coups, and revolutions. Revolution is expensive compared to evolution, but sometimes it is necessary in order to generate the changes at a pace that is acceptable. Evolutions occur over time and can often be carried out without violence.

> Even though they write the laws to allow them to be skirted, politicians still breaking those laws for a competitive edge.

As a form of low violence warfare, politics can be vicious in the extreme. Politicians assassinate characters rather than people. They lie, mischaracterize, mislead, make false promises, use false premises, send money to friends and allies, steal confidential information, and on and on. In essence, almost anything goes in politics, up to and including things that are illegal. Laws that are supposed to protect the people of democracies from unfair campaign practices are written by politicians for public consumption and are often designed to be skirted.

But even the advantage of writing the laws doesn't stop politicians from breaking them regularly. And politics can use national security infrastructure for political purposes, like spying on the opponent in the upcoming election or using secrecy to make it seem like one person knows more than the other, or revealing secrets at strategic times, or sending money to key districts prior to the election, or making it harder for certain groups of people to vote, or giving the voting machine contracts to people who have expressed the willingness to cheat to get their side to win, and so forth.

Politics is clearly a dirty business, but it has rich rewards for those who are willing to undertake a public life. Those rewards include things like getting their children into the best schools, security services for life, lots of job offers when they leave office, enormous speaker fees for companies they have helped while in office, and more... and those are the legal ones.

Politics is somewhat unique in that it has regular cycles of high and low intensity conflict associated with election cycles. While parliamentary systems can have votes of no confidence at any time, systems like that in the US have predictable cycles. These cycles are particularly useful for those in power because they can schedule their popularity to peak just before elections while becoming very unpopular and airing dirt in the inter-election periods. This sort of cyclic intensity makes these particular political systems particularly interesting because there are interesting strategies that arise as a result of timing.

Of course some political systems have only a very limited electorate. A ruling junta is typically not a democratic group, but still, if the land owners and military families don't agree with the leader, a coup will result. Even in the most hierarchical and oppressive of governments, internal political pressures force events. In theocracies, for example, even though the leaders are often considered to be directed by God, there are internal political challenges, forces have to be weighed, and there is jockeying for power, money, position, and privilege.

Politics is one of the best examples of pure information warfare that exists, and at the same time, its very existence is designed to reduce the amount of warfare in the sense of violence that takes place.

So control of the people by influencing their will and control of subsets of the people for everything ranging from funding to support to safety to prosperity are all in the realm of the information warfare that politics encompasses.

9.1 Timing

Politics has a lot of timing involved in information operations. This includes very short time frames for tactical responses to military acts, timing in the range of hours to respond to things in time for news cycles, timing in the range of weeks for scheduling bad and good news, timing in months regarding seasonal issues, timing in years regarding election cycles, and timing in decades regarding long-term dominance of court appointments and overarching political views in the society.

- **Tactical time frames and public safety and military operations:** Politicians have to deal with information operations in near-real-time when they relate to crises that affect the general public. These are generally emergency communications, communications associated with attacks and wars, and communications associated with natural disasters. They require concise and meaningful information passed to the public that reassures and causes them to act in specific ways. There are also internal communications associated with these events ranging from command and control of military operations to communication with other politicians.
- **Unscheduled news cycles:** While politicians prefer to control the news, things happen, and when they do, politicians have to deal with them. Most of these deal with leaks, investigations, foreign affairs, political and legal disputes, and of course when these crises grow as a result of corruption, they tend to be surprises with little or no notice. As the news cycles around the world become more and more rapid, near-real-time response is required in time to appeal to audiences before they become disenfranchised or before the media decides that you are non-responsive. These are often artificial and yet very important to the politician who wants to control the messages getting the constituents of various sorts.
- **Scheduled news events:** Political events are commonly staged with timing designed to produce outcomes. For

example, when a law that grants police power is about to come to a vote, news releases about the victories of this law or the failures of the old law are often scheduled to pressure others into action and get the public opinion and news media focused on the points of view desired. Similarly, things that are not to gain much attention but that have to end up in the media are scheduled for the evening of a major national vacation or some similar event so as to minimize their impact and the number of listeners likely to care about them.

- **Seasonal issues:** Economies tend to be seasonal with growing seasons in the summer and holiday seasons in the winter. The timing of these seasons leads to different sorts of news and the sound politician schedules different events for different seasons to assure that they optimize perceptions. Opposition typically talks about job losses or economic difficulties in the early part of the year (in the Northern Hemisphere) because that is when many seasonal workers are out of work.
- **Election cycles:** Election cycles are fundamental to political processes and thus there is an effort to emphasize things that will come out and are negative long before an election so as to minimize shock value while emphasizing positives just prior to elections to create peak image when it matters to the vote.
- **Dominance:** Long time frames in politics go to issues of long-term political dominance and control of the agenda in terms of what people think and care about. A party in power often seeks to optimize long time frames by creating conditions amenable to claiming victories, benefits, and so forth. Things that are controversial are started early in the cycle and settled before the next major election. Timing of dominance of specific offices allows the decision on what judges are put in to what positions, leading to long-term legal control at a philosophical level.

The list goes on in each of these areas, but you get the idea.

9.2 Divisive issues

Politicians use what is commonly called political calculus to determine where to position themselves in terms of issues. The selection of issues is fundamental to controlling the focus of attention. And the sound politician understands what portions of what populations in what regions care more or less about what issues. The calculation can then be done to determine where to stand on issues so as to win what parts of the vote.

As an example, if 10% of the population will vote for anyone who agrees with issue X and your opponent disagrees with position X, you can gain a guarantee of that 10% by supporting issue X. But because voters are so complicated and have so many views, it is very complicated to determine the proper combination of positions required to win in any specific circumstance. In addition, how things are presented, selections of language that mean something to one group and little to another group allow the politician to signal to specific constituents without offending others. This is the so called wink and grin strategy.

An approach to understanding these issues comes through the mathematical analysis of strategies called *game theory*. Game theory is a somewhat obscure field of endeavor that provides the mathematical underpinnings of optimization in various situations. I will be discussing this in some depth later, so suffice it to say for now that by using mathematics and measuring and analyzing the numbers you can pretty much figure out what messages are likely to produce what results.

Having said this, the political process is quite complex with moves made by all sides, and it is highly competitive because of all of the benefits that come with power. To get a sense of this, it is important to understand how power is used once elections are won. This is part of the information warfare that continues to get and keep people in power through strategic efforts to control views and options, and is thus central to political information warfare.

9.3 Pork

Pork barrel politics is the norm in the US and throughout much of the world as it always has been. The party in power funds projects in the places that will help them get re-elected, or at the state and local level, the districts, counties, and cities where they got funding or need votes. Put a military base up here, build another highway there, and you feed the economic well being of the folks who voted for you while systematically moving that money away from those who voted against you. It's perfectly legal.

But to discuss pork in such simplistic terms is to ignore the great subtlety involved in the proper use of pork. Pork can be indirect and subtle or direct and harsh. When direct and harsh it is easily recognized and explained creating some level of opposition that can be easily vocalized. The idea in politics it to make things you don't want people to understand hard to explain in simple enough terms to keep the voters from being influenced by them. While some portion of the population of interest will take the time to understand the issues, this can be put into the political equation and it usually comes up unimportant in a democratic society. But in other societies, this becomes harder to do because those who are voting are more knowledgeable and have more direct access to information. A military Junta for example, involves few people and they have to be pleased at great length compared to a republic in which election of legislators who make the decisions are used.

Some pork is obvious and some of the corruption that is associated with it is striking. In the US, for example specific laws are passed that exempt individuals that are friends of specific politicians from taxes, grant them money from the public funds, and even grant them land, title, position, or exemption from crimes they have committed. Leaders are often able to issue pardons thus allowing the criminals who work for them to walk free even though they have committed crimes that put others in jail for lifetimes. This legalization of crime is the ultimate pork and the ultimate power that allows a leader's workers, at their bidding, to do whatever they want and get away with it.

9.4 Taxes and land grabs

Taxes provide for the redistribution of wealth and the provision of services. The government services derived from taxes are well known and tend to include things like national defense, salaries for politicians, the creation of palaces and similar edifices of government, the manufacturing of news and other similar media content, so-called franking privileges that allow those in power to use your money to convince you to keep them in power, travel and support for speeches and other political information provision, collection of information on the people and its analysis and provisioning to the politicians, intelligence processes not related to the people in the country, the system of laws and police efforts of the state, schools and other educational and religious institutions, and so forth.

Class wars are a form of political information warfare.

When the tax collectors force you to give them money to spend on things you don't want. It's called redistribution of wealth. Of course we all tacitly agree to this by remaining here. We could go somewhere else where they tax more or less and provide more or less or different government service. The thing that gets really offensive to most tax payers is when the money goes to political cronies of the people in power, assuming we aren't among them. This goes beyond the standard pork into the realm of theft. For example, when the powerful decide to send a few billion dollars to their friends or business associates, that is over the line, whereas when they do the same thing through an official process that allows others to compete for the same funds even though the competition is largely rigged by the process, that is typically considered acceptable.

Various attempts have been used to try to use the tax system to generate voters for one side or the other, typically based on a class war approach. Whether we like it or not, class wars is what politics often ends up about because this is where you can divide the electorate in a meaningful way to assure votes of different groups.

The US Supreme Court recently ruled that eminent domain laws allow government to take land from you and give it to someone else so they can make more money with it. This is just the latest in a series of breakthroughs in the eternal land grab scheme.

In essence, politicians use land grabs to advantage their friends, to make money to support public policy, and to get even with political enemies. This is the nature of power. Whether the excuse is building a highway or increased profits for the governmental entity at whatever level, the excuse is good enough for those in power to take advantage.

> Individual land ownership for all is really a relatively new concept.

This is nothing new of course. Kings awarded land to people that did them favors and taking land from one group and giving it to another is a time honored tradition of politics. In the large this can take the form of nationalization, in which entire industries are taken over from private companies by government. In biblical times there were land owners as well, but they all derived their ownership from a higher form of government, including and up to divine providence which ultimately led to Abraham's family gaining land ownership, and in the end, this process led to the current conflicts in the Middle East. But while families owned land in some places in history, most government had kings and princes and knights and surfs and peasants for quite some time. In the end, all land belonged to the kingdom and was allocated by the king as part of power sharing to those who helped to keep the king in power.

Private land ownership came eventually but it never completely took hold except for those who were able to hold onto their land in the face of physical attack. This physical attack remains today as the way land gets transferred except in societies that have agreed to leave the physical part of the attack to the police and traded physical warfare for legal battles, another form of information warfare.

9.5 Campaign seasons

In the US, many politicians seem to come from poor families and have worked themselves up from dirt. In fact most of their personal stories are pretty close to true. But when it comes to military service, plenty of folks embellish either directly or by inference.

Think George W. Bush and the aircraft carrier where he “landed” and declared the end of major combat operations in Iraq. He was in an air force pilot's uniform as if he landed on the carrier when in fact he was a pilot in the national guard during Viet Nam but never flew onto a carrier or anywhere near combat. It's a subtle sort of lie. Not like Clinton who faced a camera on national television and told the nation that he never “had sexual relations with that woman”. Of course who knew that "sexual relations” didn't include the kind of sexual relationship he had with that same woman. It's another outright lie. And lest we forget, neither of these folks were the first US presidents to be caught in lies. And of course the President is the top politician, just imagine what other politicians are doing. I guess you really don't have to imagine. They seem to get called on lots of lies and yet they get away with many of them too.

> Campaigns are aptly named for the ferocity of the information warfare they bring out. Like the military campaigns they emulate, they are no holds barred contests in which truth has almost no relevancy and perception rules in every meaningful way.

Each of these represent the introductions to campaigns of various sorts. They were not in the heat of battle of course, they were part of the run up to elections. After all, in order to appeal to the country, it was necessary to end and win the war in Iraq, and Clinton could not lead the party if not a perfect husband. This is the nature of political campaigns in the US, and similar events happen throughout the world. Of course democracies are the most abusive of the truth because they have so many different forces to deal with that the truth is too hard to keep track of properly. Lies are expected and thus don't have to be tracked.

While these light deceptions are commonplace, far more direct lies are often used complete with the cover up language we hear from politicians who don't want to avoid questions but won't answer them either. In one classic example I heard a politician declare a complete lie that could not be challenged by anyone who knew the facts because the facts were classified. Anybody saying they were wrong would be violating the law, but they weren't violating the law by providing false information. Insane, isn't it?

Campaigns also bring out the outrageous. From radio hosts accusing the opposition of murdering people and covering it up, to the link between the opposition and the evil empire, no truth is too precious to be violated, and no technique is out of bounds for spreading rumors and lies. Anything to win, most importantly, accusing the other folks of being willing to do anything to win.

> Push polling is a classic political innovation of the information age.

Just because information warfare has been here for a long time does not mean it lacks innovation. Indeed, some of the most innovative techniques show up in modern political campaigns. With the enormous amount of research and development involved in this process, it is a small wonder that new breakthrough techniques are showing up at a rapid pace. Push polling was the result of one such real political genius (fraudster). In this scam, the supposed pollster asks a leading question like: "If Murphy admitted that he sent millions of your tax dollars to his best friend in low interest loans that were never paid back, how would you vote on his election run for governor?" The notion that it's a poll is a lie, and the premise is also a sort of a lie... “I said if...”. “Do you still beat your wife?” is old hat by now and the level of sophistication has gone way up. And the winners for 2004... "MoveOn.org" and "Switftboat Captains Against Kerry". The candidates denied involvement while cheering them on, and each told as radical a set of lies as you could come up with and each likely gleaned many hundreds of thousands of votes along the way.

9.6 Crowd controls

While the subject area is rich with other examples, I will close out the information warfare aspect of political campaigns with one of my favorite information warfare tools, the perception of crowds.

Crowds are particularly interesting things because they make great shows and can be broadcast globally to demonstrate that the radical position is not radical at all. It forms a sort of social proof that the ideas portrayed must be legitimate. The *Million Man March* is a classic example of a factor of ten exaggeration of the number of attendees at a rally. The Park Police in Washington count attendees at rallies and publish the results which are used to make claims by politicians about the crowds to counter the claims made by the other politicians about the events. And the counts are highly politicized because of their propaganda value.

> Make no mistake about it, politics is about show and the same values that drive other elements of entertainment drive the information warfare aspects of politics.

The terrorist group Web sites advise on how to generate larger crowds for the media and emphasize events that bring in large crowds by the creation of traffic jams at rush hour. Thus there is a strong desire to bomb at rush hour for maximum effect and maximum show value. If the show can be extended, all the better. In a slow news week, there can be a lot of show value and media attention to political actions, and these can drive popularity ratings and other similar measures of political situation fairly dramatically. A war always helps the leaders in the beginning, especially with proper propaganda preparation and a justification (anything will do of course, as long as properly portrayed). But if you add crowd scenes, the popularity grows rapidly and stays longer. Consider the impacts of protest rallies, even in China where the news can hardly go anywhere the government doesn't want it to go. Internal crowd numbers and their perception can drive political events like nothing else. And this is why politicians pay so much attention to the issues of crowds.

The Bush presidency in the US is one of the best recent examples of a democratic society taking advantage of crowd impacts, but even the strictest dictatorships use crowds. North Korea makes an enormous deal out of large crowds acting in concert to support their leaders. They have practically complete control over every aspect of the lives of their people and yet the crowds are required to prove their authority and right to lead. The Soviet Union had regular parades and similar events at which hundreds of thousands of people showed up to participate. How could their leadership be denied when that many people agreed with them?

> The creation and depiction of crowds is a fundamental tool of politics.

But what if you can't get a crowd or the crowd might disagree with you? That's easy to mange. You manufacture the crowd. In Iraq, crowds were regularly manufactured for rallies and political events both before and after the government changed. The supposed crowd around the toppling of the statue of Saddam Husein in Baghdad was a tight in shot that portrayed a national event that actually involved about 100 people including the US soldiers staging the event. And before the US overthrow of that government, their were enormous manufactured crowds at any event that Saddam Husein wanted to create. Spontaneous rallies in almost any land can be created, from France, to Great Brittan, to Germany, to India, to Cuba. Manufactured crowds are a fundamental of all political propaganda.

One other note on crowds before we go away. Protesters have to be controlled if any campaign is to succeed. Otherwise the consistency of support for the candidate's message will fail. This is an area where security of the politician is now used as the universal excuse to select the crowd and place protesters at a long distance. So not only must friendly crowds be created, unfriendly crowds must be suppressed.

And of course the press has to go along with these things, but more on the press later.

10 Heart Throb and other war games

> Scenarios played out to understand the effects of decisions are called games and supported by a mathematical area called game theory.

Politics is one of the most interesting and underplayed areas of information warfare and that is why *Heart Throb (HT)* was invented. *HT* is a set of scenarios that I created to explore the political issues of information warfare in a national security context. It applies full spectrum attacks to political takeover. It is now played at least once in almost every student's graduate program where I teach, and it has a spooky relationship to political overthrows around the world. Those wishing to actually take over a country can license the game from me, or I can run the game to help you get the kinks out of your plan. Or you could become a graduate student and gain access to one run-through for free, as part of your tuition-paid courses of course. Of course if you do become a graduate student and run through the game you will end up having to take the rest of the classes, and by the time you graduate you will decide its better not to take on the responsibilities of running these massive bureaucracies, but that's a different book.

The takeover of a country can be done without violence of course, or at least with very little violence. Ask the people of India if you don't believe me. But it would be far less exciting as a game if we took over using no violence whatsoever. And of course a little violence dramatically reduces the time required to take over a country and thus reduces the chances of someone else taking over as you take over. But going faster makes it more likely you will get caught. This is why you need all of the elements of a sound information warfare strategy and the tactical capabilities to get the job done without undue risk in order to succeed. That's what *HT* is all about. Of course the players aren't supposed to know the nature of what is happening before they start the game, so if you are going to attend one of my games, don't read what you have just ... forget it.

HT is an oppositional strategic 2-group non-zero sum open-ended limited repetition simultaneous move game with imperfect information. That's game-speak for:

- There are two sides that are in direct opposition to each other in terms of their goals.
- Each of the sides are teams of players.
- The game is aimed at influencing leadership over a substantial time frame well beyond that of a single battle or incident.
- When one side loses something the other side does not necessarily gain the same thing.
- There are no limits to what the players can consider or what actions they take, so they can propose anything they want.
- The game has a finite number of moves by each side.
- The teams move simultaneously rather than taking turns.
- Each team gets limited and not always accurate information about the other team's moves.

These are just some of the parameters that can be altered in games depending on what is desired by the game's creator. There are plenty of good books on game theory, so I will avoid the details for now, but the key to understanding of games is in the notion that there are large numbers of sequences of events that can take place and each participant is trying to make *moves* so as to attain their own goals by selecting moves out of the currently available possibilities in the space. Depending on the nature of the game space, situation, and participants, different moves may be more or less likely to succeed, A *strategy* is a method for picking moves as a function of situation. The goal of strategic games is to create and test strategies, which is to say, to figure out what is more and less likely to work in which situation without having to use hindsight and go through the real situation time and again. Practice makes perfect – or at least better. That's the idea at least. I'll go into more details of *HT* after we cover some of the basics of games in information warfare.

10.1 Game theoretic types and real game types

Game theory includes a wide array of different game types which are given mathematical descriptions, analyzed, and can provide optimized strategies for different situations. It turns out that in many cases the game theoretic situations accurately model real-world situations, and even more so in the information arena than in the physical arena. But these theoretical types are not as readily usable in conflict situations as those who do the mathematics may believe, because the players in real-world conflict have a tendency to be quite smart in some cases.

For example, if they see a situation in which game theory applies, they will get an expert to help them understand optimizations and work to change the game around to their advantage. An even more skilled opponent may present what appears to be a simple game, awaiting the right moment to demonstrate that it is no such thing. So deception comes into play and games like poker become interesting studies for mathematical analysis.

Typically there are several key dimensions to game theory.

- Games can be one-time or repeated depending on whether moves are played again and again.
- Games can be memoryless so that the participants learn nothing in adapting strategies or have different restrictions on memory.
- Games can have secrets or not as indicated by perfect or imperfect information for the participants.
- Games can be zero sum if my win is your loss, or non-zero sum in which case all players might be able to win.
- Games can be competitive, cooperative, or combinations thereof, so that the goal of groups may be considered more important than individual outcomes.
- Games can have anywhere from zero to an unlimited number of players.
- Games can have turns or other orderings on moves or they can be completely asynchronous.

Games of this sort are serious tools. They act as representations of conflict with the goal of finding efficient and effective strategies for winning, whatever that is. And at a most fundamental level, consistent success in information warfare or other conflict requires a good model of the situation and sound analysis against that model If the model reflects the important aspects of reality well, then it leads toward increased likelihood of success in each encounter.

> Modeling the modeling and simulation capacity of the opponent becomes a form of meta-warfare.

But of course all sides are well aware of such modeling methods and their limits, leading to the meta information war of the best information warriors, modeling and simulation warfare. Modeling and simulation play a vital role in effective warfighting and are deeply embedded in the capability of modern conflict to be favorably resolved.

In the simplest form, simulations end up being discussions or table-top exercises, and they get more and more complex, using automation of various sorts to facilitate operation of the game, automating the analysis of results, and running millions of game scenarios through a computer to understand different strategies and their effects. And since modeling and simulation are so vital to effectiveness in warfare, modeling the opponent's capacity to model and simulate is increasingly becoming a key to victory.

Modeling and simulation are used in all sorts of competitive arenas, ranging from analysis of business dealings to simulation of infrastructure prior to implementation, to modeling and simulation of networks for attack and defense, to modeling of participants in activities to understand how to influence situations. All of these are used in business on an everyday basis, if sometimes only in the heads of the participants. All can be automated, analyzed, and improved to become strategic advantages in high intensity conflict. At the top end comes operations research, which many credit with winning World War 2, along side cryptanalysis of axis codes.

10.2 Operations research and gaming

In World War 2, the US undertook an effort to use mathematical modeling to improving military performance. This applied relatively obscure mathematical principles to build up the science of warfare. The idea was that by doing analysis of situations and performing a limited number of experiments, optimizations might be found that would improve the decision-making of commanders and the selection of weapons, tactics, and strategies.

To say that this was a smashing success would be a great understatement. For example, the battle of the Atlantic involved a set of strategy and tactics options for assuring that cargo got through to England. At the time of the application of operations research techniques, the battle was being lost and supplies were in serious jeopardy of not getting to England in enough volume to sustain the war effort. But operations research looked at the numbers for different classes and combinations of combat tactics and strategies in antisubmarine warfare and came up with tactics that turned defeat into victory. This was of such import that it may quite literally have enabled the invasion of Europe from England.

Operations research is the underlying theory of optimization that is widely used in business today and includes a range of problem solving methods including:

- **Linear programming:** This is a method for finding optimal solutions for combining competing strategies to optimize overall outcomes.
- **Integer programming:** This is used for strategies that take on finite numbers of values rather than continuous ranges of values.
- **The transportation problem:** This is the well known problem of minimizing the path of a set of vehicles taking a set of supplies from a set of sources to a set of destinations.
- **Distance networks:** Shortest and longest paths and the traveling salesman problem are typical examples of trying to optimize outcomes per effort.

- **Flow networks:** This includes things like the lowest cost transshipment problem and the optimum usage of a set of manufacturing capabilities with unequal costs.
- **Queuing systems:** This includes things like optimizing lights in an area of a city for evacuations and for rush hours and determining how many people you need at cash registers to most efficiently process customers.

These are used in businesses of all sorts to gain competitive advantage. For example, in the trucking business, companies spend a lot of money trying to figure out how to optimize loads so that drivers spend as little time and gas getting as much value to customers as possible and getting the right things to the right places at the right time.

In warfare, this is used, for example, to determine the best set of guns to provide to a set of different military units given that there are a limited total number of each kind of gun and that the enemy has similar problems to solve. In the 1st Gulf war, these techniques were used to determine which set of targets to attack each day, with the result being that fewer bombs had better strategic effect in accomplishing the goals needed to win the war with the least amount of effort and loss of allied life. Large-scale simulation efforts supported these efforts along with fairly obscure mathematical analysis in the area of optimization.

Similar methods are used to minimize the cost of warfare, which is certainly a critical element of long-term national well being. How many troops should I station where for what set of enemy attacks? How many tanks do I need and where should I put them? I only have so much fuel, where should I use it first? The list of questions goes on and on, and all of the solutions lie in operations research.

At every level of conflict, the side that uses operations research has a significant advantage over competitors, and for this reason, the mathematics of gaming is vital to victory in conflict of all sorts.

10.3 Strategic games

Strategic games are used to develop long-term plans, make complex decisions, and enlighten mixes of *players*. Of course in this context a player is a participant in the game and not necessarily someone who is participating for sheer enjoyment, although of course, these games tend to be mentally stimulating and thus enjoyable to people who like that sort of thing.

The typical objectives of a strategic game are to create and explore a decision space and/or to improve player knowledge and skills. Along the way, these games help to build communication and teamwork, leading to a sort of group cohesion that comes from common experience and an ability to work better together in future situations.

Typical strategic game processes include:

- **Day after games:** Day after games come from military table top exercise sequences. They are typically played as a series of moves. Before each move, a sequence of events is described leading to the current situation in which the participants have to develop their move, which is typically a strategic approach or a set of things that different people are to do. As the series of events unfolds, participants develop and make their moves, reveal select aspects of their moves, and then get the next portion of the scenario. After the event sequence is completed, a final move is undertaken to return to the present time and rethink the initial strategy. Thus it is a "day after" game because you get to return to the start after events have unfolded to see what you can do better.
- **Prosperity games:** Prosperity games are designed to get groups of people together in order to help them develop partnerships, teamwork, and a spirit of cooperation. It helps the participants increase awareness of the needs, desires and motivations of other stakeholders and brings conflict into the open and manage it productively. It does

this predominantly by setting all of the participants at the same table and pushing out different aspects of the space to be explored in scenarios where the participants must work together to solve complex problems with urgent needs. Problem after problem is provided with the intent of giving realism to the scenario and anticipating projected problems that are likely to arise.

- **Oppositional games:** A typical example might involve well-known competitors planning long-term strategy in an emerging circumstance. Two or more groups are picked representing different parties to the conflict. They select moves by working as teams to exert their resources toward goals that they define for their teams based on their knowledge of the field and situation. Exit briefings are provided each move and the teams compete for the best and most clever ideas to defeat their enemies. Each move may also deal with a different scenario.

In each of these example types of strategic games, analysis is undertaken after the game is played and detailed write-ups are created to identify strategies, benefits, and risks. Analysis may lead to detailed operations research results for the strategies developed so that strategy mixes can be projected for making real moves.

Major limits of these games include:

- They are context bounded so that they self limit the action by the assumptions and directions associated with the moves.
- They make assumptions that limit the space explored and the exploration methods.
- They are often breadth first search instead of going into depth in understanding the real implications of the ideas.

Ten people in a room for an hour can't do the same things that one person in a room for ten hours can do because they lack deep analysis. But they do bring more ideas over a broader spectrum.

10.4 The theory of groups

The reason that these strategic games have a hope of working is because they are designed to bring together people with relevant expertise in the hope that they will enlighten each other and inform the analysis process. So the question soon arises of how to form a group for such a game so that the results are as good as they can reasonably be made to be and how to form the process so as to make the game workable in a practical sense.

The theory of groups (not the mathematical *Group Theory*) asserts that the optimal group is formed by covering the necessary and desired expertise at the level of depth necessary to address the issue while not having excessive redundancy in any area. So in forming the group for any of the strategic games described above, the group selection starts with the definition of the space to be covered and moves from there to identifying the knowledge level of the individuals. People are then sought to participate in the groups.

Of course personality traits and individual animosity come into play in group activities, so some care has to be taken not to put two enemies in the same room unless a fight is desired. Then it becomes more of an entertainment activity than a strategic scenario game.

There is a fair amount of conflict management in managing the group process and thus trained facilitators are required in order for such groups to work reasonably well. And the processes themselves have to accommodate the participants in order to have both a sound result and a reasonably pleased set of participants after the fact.

Most such scenario games are undertaken with a specific customer in mind. The customer is typically involved in the game and participation as a player or involvement of a set of players brought by the customer is done for the purpose of generating results for that client.

World War 3: We're losing it...

People who understand group processes will notice that different sorts of behaviors by individuals can upset the ability of the group to make progress. For example, by looking at the work on interactions between people, Satire's modes may be observed in these processes:

- Blaming: If matched by other players, it will lead to fights.
- Placating: If matched, it will lead to unproductive delay.
- Computing: If matched, it will lead to slow productive delay.
- Distractive: If matched it will lead to hectic mad senseless results.
- Leveling: If matched, it will lead to honesty, sometimes to excess, if you believe that this is possible.

The resulting behavioral patterns are often undesired, so it is important for those who facilitate these efforts to recognize and short circuit these problems. But this does not mean that a generic facilitator can facilitate any such effort effectively. I have seen many cases when such a facilitator has wrecked the results by deciding that all opinions were equal or that going too far into areas of particular import was a waste of time. Ultimately, the customer must be able to get involved in helping to decide what to cut off and what to let run.

When teaching students about these games, we almost always start with a lecture that I give to myself when I am about to enter such a game. It starts by explaining that this is only a game and that all of the participants must be respectful of others. That we take turns talking, and that the facilitator, if doing their job properly, will take care of keeping things reasonable when the game gets exciting and people start to lose their heads. And most importantly, in such a game, you should bring a paper and pencil (or pen) so you can write when you want to interrupt. You can then express what you wanted to say when your turn comes, politely, in a well thought out manner, and without missing the things you wanted to say.

10.5 Hearth Throb

Heart Throb (*HT*) is one such strategic scenario game played to understand the political information warfare that is only a small step more intense than current elections are around the world. The game represents a strategic attack on a country that exploits its political system and dogma to take power and make dramatic changes, all at the behest, in this case, of an outside influence that shall remain unnamed and unresolved at the end of the game.

I don't want to give the game away for those who have not played it yet, but in order to explain it well, I am afraid I will have to. So those who are interested in playing it, or those in one of my programs who has not yet played it should skip the rest of this page and the next one.

HT has two teams, an attack team and a defense team. The attack team is the council of *HT* and the defense team is the national security apparatus of the affected country – typically the US in my efforts.

HT starts with a scenario in the near future. An election is soon to be held and a suspicious series of events takes place that lead the defenders to act in the best interest of what they think is their national security, but like in the real world, they do not yet know of the existence or nature of their opponent, only that there is one out there. They are typically told of various sorts of violence that have been collated and analyzed to mean, in essence, that a group of unknown origin is willing to kill people and is preparing to attack various critical national infrastructure targets because of the intelligence efforts that the group is undertaking. The briefing includes the fact that a few specific individuals who were covertly working on the research for this correlation have recently shown up dead or missing and the intelligence on this issue is coming from unusual sources of some sort. The facts of the events are assumed to be true and we may even use recent news events as our sample data to make it all the more realistic. The question for the defense is what to do. The offense, on the other hand, is given their set of goals and asked to find ways to achieve them.

The offense, *HT*, has the goal of gaining enough political control to win a strategic victory that ultimately results in the target losing its leadership in the world for a long time to come. This is to be done by creating fear at an intensity level that causes people to vote for desired candidates, killing a few select candidates in select local districts to alter the balance of power in those districts, and using a lot of resources while unnerving the people of the country.

In move after move, the defense has striking new problems showing up that cause it to react more and more strongly and with increased national effect, while the offense intensifies its attacks in ways that continue to surprise the defense and that will ultimately cause them to look inward, increase fear, restrict travel, reduce privacy, increase the notion of a police state, and distract them from the real cause. And in every step, the defense worsens its own position by trying to respond to the attacks. Thus, this is an information warfare attack against the people running the country to cause them to use their own national security apparatus against themselves, distracting them from the real attacker while sowing the seeds of revolution. All along, the people winning positions are controlled indirectly by *HT*, and will create the backlash in the country that will focus themselves even more inward, allowing the real attacker to gain global influence and control as the defender shoots itself in the foot, elects the very people desired by the attackers, and devolves its society.

> *HT* was not specifically about terrorism and China's moves underway, but it does seem to be a decent model for what is happening today.

For those who want to know, this scenario was developed and run before September 11, 2001, and I had no idea that the Bush administration would do these very sorts of things with respect to its view of the war on terror or that China would be taking advantage of these moves as effectively as it is. Nor was China the specific country that induced the attacks in *HT* – the country was unnamed, although China was certainly a model for it.

10.6 Automatic games and simulations

Automatic games are games that play themselves. In other words, they are not designed to engage people in activities to get them to think through issues or to get people to reveal information that might be valuable for later analysis. Automatic games are also known as simulations. They are, in essence, computer programs or similar devices designed to analyze situations based on input parameters so as to yield solutions to specific questions posed about those situations. Most often, the goal is an optimal solution in some sense of optimality and the reason the games are automated is that they are both simple enough to yield to a reasonable model and complex enough to need substantial computation in order to generate useful results. The broader class of simulations are typically computer programs designed to try out a large number of possibilities in a controlled way.

> The goal of simulation is to predict the future.

Simulations are really useful tools for situations in which closed form analytical solutions to the model cannot be found and where the complexity of the real problem can be abstracted out to the point where the results are still meaningful but the computational resources required to get an answer are not excessive. This then drives to the issue of the granularity and accuracy of the representation and the future event horizon that the simulation can cover in the available time.

Simulations are, in essence, ways of trying to predict the future. Different classes of simulation technology include time-based simulation, event-driven simulation, and my personal favorite name, time-warp simulation. Time-based simulation is typically used for tracking physical phenomena such as heat flow through a material because every value has to be evaluated at every time step. The more time steps per unit of real time, the more accurate the simulation will be in terms of the time-behaviors, but the more time it will take to run the simulation. Event-driven simulation is for situations where nothing of relevance happens until some event takes place. For example, in simulating a cellular phone, there is no reason to do computations every second when the phone is not in use. Rather, one can simulate a call arriving in 15 minutes by moving time forward 15 minutes, calculating any relevant changes like reduction in available battery power for 15 minutes one time, and simulate the next action. Time warp simulation is a variation on event driven simulation that I wanted to include because it has such a cool name. It can save time on parallel processors, in many cases, through some really clever efficiencies.

The granularity and event horizon issues become clear in light of event driven simulation more than time-driven simulation.

- **Granularity:** For event-driven simulation, in the cell phone example, some might say that the calculation in 15 minutes ignores the movement from cell site to cell site. And that is exactly right. It ignores that aspect of cell phone behavior. The reason for this is that the example is based on a model that trades limited accuracy and granularity for better performance. If the goal of the simulation is to understand cell site behaviors, this simulation will not do, so we will need to model movement of the individual cell phones and use events like entering a cell site, leaving a cell site, and so forth. The simulation will take more computation for the same amount of real time but yield more useful information about the aspects of the issue we are now interested in. If battery usage accuracy is more vital, we might have to model the distance from cell sites and the power utilization implications of signal strength, the use of the cell phone by many to check the time, which turns on the backlit display for a few seconds, and so forth.
- **Event horizon:** But on the other hand, if we are simulating more accurately, the amount of available computer time will be able to look forward in time a shorter distance per unit of real time because more computations are required for the more accurate representations of reality.

10.7 Situation anticipation and constraint

An area of particular interest for me is model-based situation anticipation and constraint. The idea is to create a model that can be simulated with a desired degree of granularity for a desired amount of future prediction over a desired amount of real time and to run that model on many scenarios so as to provide limited coverage of the space of possible future events within some time window into the future – up to the event horizon of the simulation. Based on moves of the parties simulated, outcomes are measured in terms of some metrics, with the idea of measuring the best and worst scores at the event horizon, the so called minimum and maximum values, associated with every move I can make today. I can then make moves today that offer the best tradeoff between risk and reward be selecting those with the proper mix of minima and maxima.

> Conflict is often better managed by constraining rather than optimizing.

Of course, this approach is problematic in many ways, starting with the creation of a model or models, and moving through to the decision about the criteria for selection of a move once I have the outcomes. Complexity arises when I consider, for example, the problem of uncommon objectives. Suppose another player's gain is not necessarily my loss. How then do I deal with making decisions about their gains? I might be seeking a win-win situation in which case it might be more valuable for me to help the other player than to gain for myself, under the theory that beyond the event horizon of the simulation, good things will come. Many people use underlying philosophies to guide these decisions and, of course, without well defined goals that can be codified in terms of the simulation, it becomes unclear what to even measure and simulate.

In conflict situations, it is often helpful to constrain the future situations rather than to try to strictly optimize them. Assuming we can measure any given situation relative to the containment objective, the objective can then be considered in light of this goal as a primary target of move selection.

In this approach, the future situations are characterized in terms of desired and undesired future situations and moves are considered in terms of the potentials for resulting in undesired situations. The goal is to constrain futures to desired situations. One of the advantages of this approach is that as move sequences produce cases that go out of the desired future at that time window into the future, they can be eliminated from further consideration. This means that as the simulation is run, it eliminates large numbers of move sequences from ever having to be simulated, allowing a longer event horizon and finer granularity for the same available computational resources.

There are, of course problems with all approaches. In this case, one of the problems is that it is highly risk averse with regard to the allowable futures and it therefore refuses to allow for risky moves with high rewards or temporarily unacceptable situations which might result in better futures. Of course these are the very reasons I like it. The fact that a nuclear war has the chance of producing a very underpopulated planet with enormous available resources for the small number of survivors brings me little comfort.

> Today we have a hot war between nations states in the strategic thought arena.

Assuming that more than one side embraces this approach, it is likely that the side that is able to create the most available computing resources and develop the best models will have a significant advantage in the conflict. We then have a next level of information warfare, the war for dominance of the strategic thought space. And indeed that is exactly the war that we have today, only the simulation engines are the people making strategic decisions and spending their time on strategic efforts.

The hot war today is the strategic simulation meta-war. It is a war for intellectual prowess in which the country that can build the best intellectual capital and most efficiently apply it will win the strategic conflict for a long time to come.

10.8 Gaming for other purposes

Of course the strategic simulation meta-war is not limited to international conflict between nation-states.

- **Political organizations** use gaming and simulation methodologies to gain advantage as well. They try out a variety of strategies internally in order to develop the best strategies they can, and the better they game them out the better their chances of victory are.
- **Tactical war gaming** has been around for a long time and has been used for reenactments of key battles in board games and to explore possibilities from planning of specific battlefield maneuvers to the development of the officer corps.
- **Strategic war gaming** has been used for everything from developing the policy of *Mutually Assured Destruction* that many say saved the world from nuclear war for more than 50 years, to planning global allocations of resources for national defense.
- **Gaming the system** is common practice all over the world. No matter what system you work in, you are always in the role of finding ways to have the system act for the goals you want to achieve. Since the systems are often not intended to help you achieve these goals, you probably consider many possibilities for sequences of things that will help you accomplish those goals, thus finding ways to beat the system.
- **Business strategy games** are used in a wide range of forms to help businesses develop long-term plans. They are widely used in much of the world but less and less in the US because of the increased focus on short-term goals over all strategic objectives.
- **Security gaming** is used to develop and roll out policies for corporations. In this mode, games are developed to determine policy decisions, played with top management to create scores for move options, then played with workers with scoring that trains them in desired moves.

- **Security attack and defense simulations** are used for analysis of security architectures. For example, there is a free single-run security attack and defense simulator on my Web site, all.net, that can be used to get a sense of the detail level at which these simulations can be run. A related paper on that same site discusses how this simulator is run millions of times for comparisons of different security architectures.

The common thread among all of these simulation approaches is their relationship to model based situation anticipation and constraint. They exist to allow their users to look at different futures.

For a better look at the future, these simulations are often fused together with group processes. In this approach, a group process, such as those described earlier, is used to create the range of significant options to be simulated and the parameters and granularities associated with understanding those options. Then simulations, or mathematical analysis combined with simulations, are used to work through large numbers of possibilities and develop viable strategies. Leadership then takes the results of this effort and applies it to situations to act on those strategies.

Of course this can be taken too far. In the reality of the world, people adapt and people are far smarter than computers at making strategic decisions, even in the best of simulation environments. One excellent example of a spectacular failure associated with the simulation approach taken to extremes was a financial services firm founded by a Nobel laureate in economics, a well known Wall-street trading expert, and a businessman. It was funded to the tune of many billions of dollars by investors they convinced to let them use simulation and analysis engines to win in the market. And they did win – to the tune of 40% return on investment for the first few years. But the market caught up with them and, before long, the advantages they had started to wither. Then things got even worse. World events went outside of the realm of what their models were originally designed to deal with. A trillion or so dollars later...

10.9 Gaming for military applications

Games are used to test out military options in much the same way as they are used for non-military situations, except of course that there is more to lose in most military conflicts than in most civilian conflicts, which means that it is worth the time and effort to do the sort of strategic work required to run war games. Another factor is that military operations are usually in preparation for war rather than in fighting wars. As a result, they have time to undertake strategic thinking, or rather, they decide that strategic thinking is more worth the time available than other things.

> Training games no longer present just generic situations. They now present very precise models of the actual combat situation about to take place.

Gaming in the military ranges from the sort of table top exercises described earlier to the simulation efforts that help to optimize and run through large numbers of possibilities automatically, finding cases where one strategy dominates another. Some of the most interesting efforts, however, use simulation environments for a very different sort of game. The military exercise.

Simulations for military practice are used to hone very specific skills ranging from generic practice between large numbers of combatants in a broad range of tactical situations to practice for specific actions against specific targets. A good example is the US military simulation system that allows the specific terrain and set of enemy combatants present in a specific situation to be played again and again by the actual attack force to get the attack force specifically trained for the actual mission. The days of climbing generic hills or cliffs to practice for a mission have yielded in high technology military organizations to climbing a wall that is pretty close to identical to the one being climbed in the actual mission, entering the same building with the same room layout and placement of guards, and so forth.

10.10 Information warfare in war games

There are two interesting aspects of information warfare associated with war games. One is the use of information warfare methods during war games to practice operations when information warfare is being used against you. The other is the use of war games to assert reflexive control over warfighters.

- **Attack on infrastructure during war games** is rarely used because, as one commander indicated: "it ruins the whole war". The problem is that the purpose of practice is to perfect techniques, but when everything goes wrong, those techniques don't get practiced. On the other hand, if the enemy in the field attacks the information infrastructure and you haven't practiced how to operate under these conditions, you could be defeated in battle.
- **Attack on the war game system to train the enemy** is a completely different approach. By causing the war games to exercise certain behaviors in certain circumstances, it is feasible to cause the enemy to act under pressure in predictable ways that allows you to defeat them.

Realistic attacks against information infrastructure by the enemy are a serious concern for highly automated war fighting strategy and tactics systems, such as those used by the US and Western European nations. But it does not do a very good job of training to have fighters defeated again and again and feel confused in their jobs.

In more direct information warfare war games performed in the late 1990s and early 2000's and involving computer network attack and defense experiments, I showed that long-term psychological effects on the fighters can cause them to be ineffective at offense after they are defeated by deception-related defenses. These defenses are specifically designed to cause the attackers to have cognitive problems in attacks, and the effects last a long time. I am unaware of any studies that assert the same results for other situations.

10.11 Military information warfare war games

I reserve a special spot for military war games specifically directed toward information warfare. The example given above of psychological effects represent one of the many dangers associated with war games involving information warfare. There are many others. For example, because information warfare involves deceptions and some of those deceptions become ineffective after their demonstration, by playing them out, their effect is defeated. This is good for the defender but bad for the attacker. So this brings to light what is commonly called the equities problem.

The equities problem has to do with the fact that the capacity to attack is reduced when the capacity to defend is increased. So the decision makers have to decide which is more important; the ability to attack the enemy or the ability to defend yourself. Of course those who are defenders always feel as if they are getting the short end of the stick, which spending figures show that they are. Meanwhile, attackers are maddened by defenders that understand how to attack because they think many of their attack methods are big secrets. I have been amazed to go to classified meetings in which widely published vulnerabilities were presented as classified. Very strange.

So at any rate, war games involving information warfare in the sense of computer network attack are particularly interesting to me. This ranges from games to practicing all facets of information warfare, the so-called full spectrum attack and defense space. This means that the efforts range from the basics of propaganda to battlefield psychological operations, sweating information out of prisoners, technical surveillance, deception operations, countering enemy information operations, operations security, trying to enhance or defeat tempo enhancement methods, feeding false information, political attacks to kill the will of the opponent's support at home, injection of malicious packets into networks, sale of Trojan horse hardware, tracking their citizenry and yours to find links between people and places, targeting with intelligence gathering and analysis systems, and so on. Sounds fun to me...

10.12 Business information war games

But even more fun are the sorts of war games that I lead in security assessments of businesses. In this case, we typically run a set of offensive experiments and intelligence gathering and analysis functions to attack the business without doing any harm. We provide the results to the client rather than a competitor or other attacker along with explanations of how the information can be exploited against them. This is a form of full spectrum information warfare directed at a specific target, identical in many ways to the sort of attacks I would expect to be made on a larger scale as part of a national attack on another nation, by a religious information warfare operation as a result of a crusade or fatwa, by a criminal organization trying to gain advantage, or from other similar threats relevant to the enterprise.

These typically involve the full spectrum of activities, but under close supervision of the client company, and with some considerable restrictions on what who can do, when they can do it, and so forth. The goal in this case is not to simulate an attack, but rather to simulate all attacks by gathering and exploiting different information in different ways and then using analysis to put it all together.

The basic notion here is that the space of attack mechanisms is explored at some level of depth and associated with threat types that can be identified for the enterprise. This is done starting at the outside with light intelligence gathering, followed by increasingly internalized attack methodologies starting with different sorts of access and using increasingly rigorous and noisy methods. While we almost never get to the level of denial of services or actual exploitation because the defenders tell us when to stop, the efforts are usually comprehensive enough to form sets of feasible attack graphs that can be exploited to large effect. These exploits are then used as examples of what can and cannot be done and the customer has the material information to back up assertions of weaknesses or to tell them where they are not vulnerable.

11 The spectrum

The electromagnetic spectrum as well as sonic and time domains afford a lot of potential for exploitation in what is sometimes called electronic warfare, but is more simple to think of as low level information warfare. In essence, information operates at all levels from the lowest levels of physics where information is intimately tied to the very heart of atomic particles and how they work through the signals level wherein communications and storage of signals are used to encode content and apply it, through the linguistic level where content takes the form of defined syntax and semantics, through the levels of behavioral detection and response, and all the way to the level of human, animal, and automated thought processes, and presumably beyond even these.

> Theoretical and practical understanding of fields, waves, and the theories of electromagnetic systems can be very useful in information warfare.

Because, in general, gravitational effects exist at arbitrary distance across the entire universe, it is, in theory, possible to derive information at any point in the universe about the situation in any other point in the universe, with delays associated with the speed of light. So, again in theory, there is nothing that can ever be done to perfectly prevent anyone anywhere from knowing anything anywhere else a very short time later. But in practice, the world does not work as well as it does in theory, or rather, there are advanced theories that tell us more about how the world works, including the limits of the ability to actually derive this information and the practical limits associated with mechanisms that we use to do so.

I don't want to go into general relativity and the field equations here, and you must trust me that it would be of little benefit to you or me, but trust me that if you really want to understand all of the low-level aspects of information warfare, you will need that as background. Since we don't have it, I will summarize the issues that are most obvious and the interested reader can drill down on their own in any good electrical engineering graduate school.

The point of pointing this out is that I am now getting into the highly technical part of the book and there is no avoiding some amount of technical information in a field that is obscure to most people. I will try to be gentle however.

The basic notion that wave forms are required for storage and manipulation of information content and that it is impossible to perfectly assure that those wave forms are under your control, means that there is a wide range of potential for exploitation. This in turn makes a capability to control the spectrum valuable in conflict.

> The equities issue rears its ugly head again in electronic warfare.

From an offensive standpoint, weapons have been developed to allow corruption, denial, and leakage of electromagnetic signals at a distance. From a defensive standpoint, there are methods that allow the defender to make the offensive methods a lot harder to accomplish. But defense is typically harder to do because the attacker may attack anything using any capability while the defender may have to defend against a lot of things and may miss the thing the offense came up with. Thus secrecy of the offense allows defeat of the defense, while a good offense allows the defense to know what the other side's offense is. Hence, even if there is the potential for conflict, the defense must press the offense to get it the information it needs to defend while the offense must not provide too much information to the defense or the defense will be able to defeat the offense.

At a more practical level, all defenses have holes but they can often be reduced if we understand them. Meanwhile, offenses are highly susceptible to deceptions but defenders are typically not as good as they should be at using them, because of cultural issues. Once attackers are detected, they can be eliminated to limit their attempts. This is the place the defenders should focus their information warfare efforts if they want to change their equities in their favor.

11.1 Wave forms

In general, electromagnetic or sonic disturbances take the form of waves. The form of those waves, the manner in which they rise and fall with time, has everything to do with the information they carry and their effect on the world around them. The folks that work on electronic warfare spend their time developing mechanisms to sense, create, and alter these wave forms to advantage. For example:

- A wave form could cause the power supply of specific types of computers to fail without affecting other similar types.
- A wave form could cause a display to become over charged and need to be degaussed, producing a few seconds to minutes of lost utility.
- A wave form could cause speakers to produce harsh sounds resulting in listeners being distracted or even having their ears affected for a period of time.
- A wave form could cause printers to seize up (my printer experiences this a lot)
- A wave form could be used to open or close a garage door without the owner pressing the button.
- A wave form could be used to break starters or alternators on cars, or even to stop cars in their tracks by disabling their internal electrical systems.
- A wave form could cause wireless systems to fail over areas controlled by the attacker while their radio equipment continues to operate.
- A wave form could cause cellular telephones to fail.
- A wave form could be used to cause radio frequency identification (RFID) tags to fail or identify themselves.
- A wave form could be used to alter signals between systems so as to cause remote systems to report the wrong data.
- A wave form could be used to change your television channel from your neighbor's house or across the street.

- A wave form could be used to cause your cellular telephone to turn on and start transmitting whatever is said in its presence.
- A wave form could be used to cause all of the pagers in an area to go off at the same time.
- A wave form could be used to cause location systems on airplanes and cars to go awry and report the wrong locations.
- A wave form could be used to light up a target for aiming a missile at it.
- A wave form could be used to prevent sounds from passing a barrier or to induce other sounds at the barrier.
- A wave form could be used to order audio inputs to computers to take actions even though the user could not hear the commands being made.
- A wave form could be used to detect the presence or absence of materials on or within people passing a barrier.
- A wave form could be used to detect movement in an area.
- A wave form could be used to detect changes to wiring or attempts to add external wire tapping devices to a wire.
- A wave form could be used to cause troops to become temporarily blinded in a military situation.
- A wave form could be used to cause intelligence systems and sensors to target the wrong locations.
- A wave form could be used to take over control of remotely operated vehicles.
- A wave form could be used to detect how fast vehicles are traveling or to counter devices that do that.
- A wave form could be used to cause groups of people to have to immediately go to the bathroom, feel sick, throw up, have intense skin pain, or become disabled.

Every one of the items I have described above has been done in the real world and is part of either a commercial or military capability in use today. And this is only the beginning.

11.2 EMP weapons

Electromagnetic pulse (EMP) weapons are sexy in the sense of having gotten some public interest as a result of their effect on normal every day folks when atomic weapons testing was underway and subsequence hyperbole surrounding their potential use in other venues. Of course EMP weapons are real and do exist, but understanding them requires a bit more than just fear.

> EMP resulting from nuclear weapons use is one of the side effects of these weapons that were not anticipated by the original designers.

EMP effects of nuclear weapons were discovered when nuclear testing in the atmosphere wiped out power in a substantial land area in the 1950s. At that time a fairly intensive research effort was undertaken to understand the issue both from an offensive and a defensive standpoint.

The concept behind EMP is that a pulse of the right magnitude and rise and fall time will cause many devices to fail, including most radios, computers, storage media, power, and communications systems. Thus, an EMP weapon of sufficient magnitude could wipe out most of the technological information capabilities of an opponent over some areas of space for some period of time.

The Russians are best known for developments in this area during the cold war where they created weapons that could be used in relatively small areas such as battlefields. This was a strong counter to the US increased use of technology in their weapons systems.

Since that time EMP weapons are rumored to have been used in many other contexts, ranging from causing outages in banks to wiping out all of the data in computers at an abortion clinic from the parking lot. Most of these rumors are just that. EMP weapons must produce a very high level of energy in order to destroy computers because they are relatively gross in terms of their wave form design.

11.3 Taking out swaths of the Earth

A very different approach to disruption of information systems stems from an effort to understand the field lines of the Earth. The Earth has electromagnetically charged poles that cause magnetic compasses to work. These fields generally run from magnetic pole to magnetic pole over the entire Earth and, in addition to protecting the Earth from Solar flares and other similar outer space effects, these poles change over time because they are induced by the electromagnetic currents arising out of the hot metallic components of the Earth's core. Every once in a while they even flip so that North and South magnetic poles change.

> Energy weapons taking advantage of Earth's fields may achieve enormous changes to the way many systems work. But the side effects may be large as well and the ability to control effects is vital to their use.

It also turns out that these fields surrounding Earth can be altered by the systematic induction of energy near where they enter the surface of the Earth. Since Alaska has access to many of these locations, the US government has done experiments and developed installations to use positive feedback in the electromagnetic field lines. Apparently, it is possible to change the underlying electromagnetic parameters for a substantial swath of the Earth by this sort of activity, causing essentially all electromagnetic functions in those areas to act quite differently than they normally do, including biological functions.

Such an attack could have devastating effects on any country that is highly dependent on information technology but there is substantial question about whether it can be selectively targeted tightly enough to be a useful weapon.

Extensions of the scientific results from this sort of research and development to energy weapons, the required energies, and how to produce and direct them are likely to be highly applicable to other weapons systems.

11.4 Tempest

Tempest is a US military term that has become widely adopted for describing emanations security. I will be discussing the broader issue of limiting the introduction or emanation of signals associated with content. The EMP weapons and similar wave form approaches are often used to disrupt or disable systems, but in the more insidious approach, wave forms are injected or examined to produce content, the stuff that has utility in information systems and technologies.

There are a wide range of methods for producing signals that result in comprehension by the systems they are directed toward. For example, by using the proper frequencies and projecting sound waves, a person can quite literally be made to hear voices coming from within their head. Similarly, frequencies that dogs hear but people don't can be used to *command* animals without people suspecting there is even communication underway. High frequency sound can sometimes be used to command computers with sonic inputs as well. Similar approaches allow surreptitious introduction of control signals into wireless systems such as Bluetooth interfaces to computers.

Tempest also involves the ability to listen to content from afar. So-called van Eck bugging is an approach published (then removed from printed copies - but some still got through) in *Computers and Security* in the 1980s. It describes how a simple television tuner could be tuned to the proper frequency and observe the content of a distant computer display because the computer display emanated signals from it's high powered cathode ray tube indicating what was being displayed. This was demonstrated from a panel van that was able to show what was on the screens of New Scotland Yard in London. It made quite a splash that police computer access could be observed from outside of the building, but the military folks of the world were well aware of these issues before the van Eck demonstration. For a long time they had been concerned about these sorts of emanations compromising national security secrets.

It turns out that it's quite difficult to stop emanations because they come in so many forms from so many places in so many ways. For example:

- A researcher recently demonstrated that the visible light from a display screen (all types) reflects off of glossy wall paints over several bounces and can be detected and used to reconstruct what was on the screen, even around corners, through windows, and at a substantial distance.
- Another researcher demonstrated that by inducing specific graphics on a screen, the display could be made to emanate AM radio signals of the display generator's choice.
- Many researchers have shown that power supplies allow high frequency information to pass through them that results in picking up instructions and data from computer central processor units and bus activities.
- Different sounds have been detected from different keys of a keyboard allowing a carefully used listening device to detect what is being typed by the different sounds.
- Timing of keystrokes also provide information on what is being typed to the point where it has been used to extract passwords from timing information alone.
- Conversations can be picked up from shining laser beams at window glass at long distances and observing the phase differences in returned light, which correspond to the glass movements resulting from the audio waves in the room.
- Many displays emit sounds at high frequency associated with what is on the screen and these sounds may be reassembled into meaningful signals.
- People mumble to themselves and careful listening with directional microphones can pick up what people "*say to themselves*" as they think through things.

There are many other examples of Tempest releases that affect people, computers, and systems of all sorts.

11.5 Countering Tempest

The problem from a defensive standpoint is how to counter tempest attack methods. There are basically three things you can ever do at a generic level to defeat tempest attacks:

- You can suppress the emanations of signals.
- You can increase the distance between the source of signals and their capture point.
- You can introduce false signals to make it harder to understand what is sensed.

Suppressing emanations has theoretical limits. Reducing power levels of sources help a lot, as does the use of a Faraday cage or properly absorbent materials. A Faraday cage is a wire mesh cage. According to the wave nature of electromagnetic phenomena, if the frequency is such that the wave length is greater than the mesh size, the waves will not pass through the mesh. But computer signals in particular use square waves which are composed of large numbers of sin waves of different frequencies in different proportions. The higher frequency harmonics tend to get through Faraday cages that have enough mesh size to allow air through. So if a full enclosure is used, it has to either enclose the air the people breath and the power needed to operate, or there will be emanations at some frequency through some channel. Absorbent material is used to reduce emissions in wireless networks to substantial effect, but this is far more effective at directing waves and limiting interference of primary signals than for reducing emanations. It is also used in things like stealth aircraft and so forth for similar effect.

Distance in the form of perimeters can be used if it can be assured that the listening devices are outside the perimeter. This then begs the question of Trojan horse hardware and its use to listen to emanations and retransmit via covert channels. Generally, the available signal reduces as distance increases and in unrestricted space with an electromagnetic signal emitted from a point source, this reduction in available energy for detection for a given sized detector goes down as the square of the distance from the source. For different shaped emitters, different spaces, and different types of signals, the reduction may be more or less. For example, material that reflects or absorbs energy in the wavelengths of interest will reduce the signal at a point, while a laser or wave guide can keep energies higher at longer distances in the direction of the path. Radio emissions at very low power, for example, can reflect off of the ionosphere, ionized layers, or different air density regions and bounce across the World while reception within only a few miles of the same source may be impossible for the same receiver and transmitter.

Introduced signals include such things as noise generators and false information interlaced with real information, frequency hopping with noise injection, and similar techniques, all of which make picking the true signal out from false signals harder for the receiver. For example, for people trying to listen to sounds, a set of recordings of thousands of people at parties having conversations could be introduced into a perimeter area so that the desired set of voices become far harder to discriminate. Similar signal injection can be used for signals from computers and other media, however, this is not an easy task to do well even for the highly educated and experienced among us. For example, the Russians introduced a piece of equipment called "*The Thing*" in a gift to the US embassy in Moscow. It was a seal of the US with an embedded cavity with a metal rod protruding into it, all concealed within the wood of the gift. It turned out that by transmitting microwaves into the embassy, the device would produce different returns based on the sounds in the room vibrating the rod within the space, and the Russians could listen to the discussions underway. It also turns out that with a predictive receiver that picks out the number of available messages from a select number of messages being sought, very high quality of reception can be done. Speech patterns of individuals may be put into a predictive receiver to detect their speech, even from within a crowd. And directionality can be used to dramatically reduce noise levels. It's all part of the tempest world of the information warrior.

11.6 Deceptions

The introduction of false signals begs the question of deception in those signals as well as elsewhere. While the general topic of deception is clearly embedded throughout information warfare, the use of deceptions in the direct analysis and injections associated with the spectrum is somewhat more limited today. In general, while everything has a representation in the electromagnetic spectrum, including all matter and energy, the complexity of generating arbitrary sets of waveforms in an area of space is, at least for now, way beyond any foreseeable future.

> As deceptions become more accurate, higher fidelity, and more complex, they also become far more expensive and harder to do.

On the other hand, the quality of the deception and its effectiveness are driven largely by the fidelity with which the deception is carried out. A cocktail party sound effect may say "*go away*" to a listener, but a serious attacker may also become all the more determined to isolate the voices of the subjects of the surveillance. A more serious defender may realize that this could happen and create far more elaborate deceptions involving the transmission of realistic information content so that even once received and analyzed, it becomes impossible to tell the fake secrets from the real ones. The use of special words and codes in communications may also augment the counter-surveillance effort to make the utility of signals intelligence far more limited.

Defenders can create fictitious targets for attack so that the incoming signals appear to suppress an activity when they only really drive it underground and give warnings to the defenders of the presence of attackers. All of this, of course depends on the capacity to create realistic deceptions at a level of quality such that the attacker is unable to tell the deceptions from the realities, even as they actively attack the overall system using the spectrum as only part of their overall effort.

11.7 Sounds and silence

One of the most interesting developments in the sonic spectrum of late comes in the form of my Boze headphones. These are active noise cancellation headphones that listen to the outside noise and other sounds, analyze those wave forms, and compensate for them be generating their own counter wave forms that just cancel out the incoming wave forms at the ear, so that when you wear them the sounds do not get to your ears. On airliners, this dramatically reduces much of the unpleasant nature of the experience and makes listening to quiet music at high fidelity possible where it was not before.

The technique is so good that the airliners even use it on the engines of their planes in some cases to cancel out the sounds of the engines for the passengers and those on the ground. This produces quieter engines and similar technologies may even reduce turbulence for smoother and less expensive flying.

> Cancellation is limited in electromagnetic systems by the speed of light.

If it works for airplanes, why not use it for meeting rooms? You can, of course, and it works well for the sounds in a proper environment However; this approach does not work for electromagnetic systems. The reason is simple really. Electromagnetic phenomena happen at the speed of light, and the electronic analysis is impossible to do faster than the speed of light, so the cancellation can never keep up with the signals unless the signals are known in advance. In the sonic world, sound travels at only 300 meters per second, so a lot of calculations can be done between the time a sound wave reaches a sensor at the edge of the headphone and when the same sound wave reaches the speaker inside the headphone a quarter of an inch away. At 300 meters per second, a millimeter is 1/300,000th of a second, while a computer can compute at a rate of billions of computations per second, or perform more than 3,000 computations between the microphone and the speaker.

11.8 Covert channels

Most of the spectrum issues I have discussed end up being important because they create covert channels for information flow. These examples have been accidental covert channels that the defender wishes they could eliminate and the attacker takes advantage of because they happen to be present. But there are a lot of covert channels also available for those who wish to intentionally induce them. All it takes is the ability to plant some hardware or software of your own design within a system, and you can readily create all sorts of covert channels that intentionally use signaling systems to transmit the information accessible by the hardware device to the outside world.

> Covert channels are any ways to communicate that are not fully announced and explained to all parties involved.

Any time a resource is shared by parties that should not be allowed to communicate, it can be exploited for communications. When this shared resources is used without explicit notice that it is a communications media or when the communications method in all of its detail is not specified, there are covert channels; channels that are not overt, or announced. The includes the electromagnetic spectrum, the sound spectrum, and the signaling protocols and paths used by these spectra to communicate.

For example, in network traffic, covert channels are available even in the most secure of operating systems when the network is shared. These covert channels include all of the variations in valid packet headers, settings, and fragmentations, time to live settings, source ports in sessions, packet timings, packet sequencing, and error responses.

This example nicely points out several of the key issues here. It turns out that in addition to the self-signaling nature of the content passed in the syntax of the communications protocol, all of the variations remaining in the protocol, including all unspecified or underspecified components are potentially covert channels.

If the protocol specifies power in the range of *a-to-b* then variations from a to b can be intentionally generated as a covert signaling method. For example, lower power can be used to indicate a '0' and higher power to indicate a '1'. Or more values can be encoded in the range from *a* to *b*. Similarly, the timing of signals and delay characteristics of responses can be varied. For example, sending out a response in less than a microsecond for a '0' and more than 2 microseconds for a '1'. And just like power levels, timing can range over a wider variation for more values to be encoded. The content of packets in a network typically include a variety of header options and value that can be modified en route and even returned to their original or suitable values later on, so that a pair of devices inside an installation and outside an installation can be used to encode covert information that only runs between the two devices and the intervening infrastructure and does not impact anything at the two ends.

With a bit of thought a lot of variations on these themes arise. For example, because an efficient algorithm optimizes operations so that results come as soon as they are available, the same efficient algorithm will produce faster results for some operations when there is a '1' than a '0' bit in any given location in the content being processed. By simply examining timing for operations, covert information about the '1's and '0's being processed may be gleaned. This has been used to decode passwords sent in encrypted packets and to break cryptographic systems by figuring out the keys when they are used to encode known content. Responses to login prompts with different timing for valid and invalid user identities and partial passwords have produced similar results. The presence of large numbers of pizza orders when a military attack is pending is a strong indicator at the Pentagon for the attack to start very soon.

Steganography intentionally uses the allowed variations in content to encode data. For example, in pictures, minor color changes are not noticeable by the viewer but can encode large messages surreptitiously.

12 Information attack tactics

Many people who discuss information war are in fact discussing the subfield that is now commonly called computer network attack (CNA). This is a fairly narrow field that gets a lot of attention in the media, but that is obviously only a small part of the attack space. Many of the people who work in the computer network attack area don't recognize how computer network attack fits into the bigger picture of full spectrum information attack and as a result, they tend to think in terms of single-step attacks rather than sequences of events that produce the desired results. In full spectrum information attack, computer network attack is used as an element in a larger strategy to gain desired effects. The typical issues in attack include:

- **General approach:** The general approaches most often used include outside in, inside out, and networked or middle out approaches.
- **Direct attack on computers over networks:** This involves sending signals into computers and generating responses. It generally breaks down into distant, proximate, and enveloped methods.
- **Perception management approaches:** These are approaches in which systems or people are given to believe things that result in desired behaviors. This includes elicitations and some sorts of deceptions.
- **Indirect intelligence gathering:** This includes all attempts to gather information on the target without any direct contact with the target. For example, looking them up in the library would be an indirect intelligence effort.
- **Direct intelligence gathering:** This includes direct attempts to gather intelligence on the target by interacting with them. For example, calling up and asking for a catalog.
- **Garbage collection and other physical intelligence:** Generally, waste products are less well cared for than the same material is cared for before being put in the waste basket, and thus dumpster diving is a popular sport among attackers. Other physical intelligence also goes.

- **Physical entries and appearances:** In many cases members of the attacking side show up at the target sites and do direct physical attacks to gain entry. This is usually surreptitious in information warfare.
- **Trojans and plants:** Via one delivery mechanism or another, Trojan horse hardware and software, planted human operatives, and other corruptions are engaged in.
- **Combinations and sequences:** These things can be combined and sequenced so as to produce desired effects when and where desired.

These are typical combinations used by reasonably sophisticated attackers that are realistic threats to all defenders in the information warfare arena. In addition to the mix of techniques, there is generally a pattern of behavior that attackers follow. The pattern is rather simple in its basic form:

- **Gather intelligence:** This involves all efforts to get information on the target that can be exploited either directly for the objectives of the effort or indirectly for gaining those objectives.
- **Gain entry:** This involves getting a foothold somewhere on the path from where you are to where you want to be, either logically via computer entry or physically by getting on the next step of your path to success.
- **Exploit privileges gained:** Entry grants privileges of one form or another. Physical entry may grant visibility, information entry may grant capabilities to alter content, and so forth. Exploitation is gaining desired objectives either for this step of the attack process or for the overall objectives of the effort.
- **Expand privileges:** Using the new capabilities granted by the entry, additional attacks may be attempted to get to new places from the old places.

At every step, previous steps can be augmented and the process occurs simultaneously and recursively over time in the attack.

12.1 Approaches and attack graphs

Non-trivial attacks are based on the notion of taking multiple steps toward attaining goals. The combination of these sets of steps produce sets of paths from the source of the attacks to the destinations of those attacks. The set of all paths is typically characterized as an *attack graph* and consists of a set of nodes that are logical places in the attack sequences and links between those places that represent different ways of reaching those logical places.

For example, in order to gain the objective of economic advantage over a company, I might want to get the details of what they pay their suppliers and what they charge their customers so I can negotiate better deals with suppliers and undercut their prices. Suppose this demands that I gain ongoing access to their internal bidding content so I can track individual bids that are competitive with mine and that I gain access to what they pay by any of a host of locations where portions of that information resides. I can't realistically just walk up to the front door and ask for these, and if I launch an attack on their Web site it is unlikely that I will get the information that I need there. Rather, I will need a more complex plan that involves gaining access to the desired information at the desired time without the target's knowledge.

At the extremes, approaches to targets include outside in and inside out. All other approaches are essentially networked approaches in which sets of capabilities are put in place and they interact with each other to provide paths through the attack graph from source to destination.

The outside-in approach is to start attacking at the perimeter and work your way into the core of the parts of the enterprise that have the information you need. From there, the information is then extracted back along the path of entry or through some other means available from there. This might involve an external physical penetration followed by planting a device on their network which is then remotely exploited via a wireless link to attack internal systems, eventually gaining the desired content and sending it out.

The inside out approach is to start by planting or gaining resources inside the target. For example, if the target is advertising for a sales representative, the attacker can submit a range of resumes under different names and try to plant an intelligence operative in one of the positions. This individual can then use their inside access to get at the system or information desired more directly and find ways to work it back out without getting caught. They might even get on the bidding committee for the work of interest and have access to pricing information as part of the bidding process.

The middle-out networked approach provides more attack graphs to the attacker while affording redundancy against defenders defeating individual attacks.

The middle out, or networked approach is more successful in most cases when there are substantial resources available. In this approach, a set of capabilities are planted against the target and involve whatever can be easily planted wherever it can be planted along with things that are harder to get in placed in locations where they are critically needed in order to gain the objective. These resources network with each other to form a set of overall attack graphs that provide paths from attacker to target and back. When one fails, alternatives are available, and of course each can be leveraged to add new capabilities.

In most of the efforts we use against corporations, we do a set of independent demonstrations and posit attackers and capabilities planted in a variety of places with different goals. This is far more effective at understanding the nature of the protection situation within the enterprise and is a far more realistic approach to understanding the protection program and its limits against real attackers than the processes we often see from tool-based solutions.

In studies of actual attacks launched against high valued targets, we rarely find an example of a successful attacker using a single strategy or a linear approach. Combined attacks are the norm.

12.2 Direct attack on computers over networks

Of course combined attacks involve many individual attacks, and these individual attacks are commonly directed against information infrastructure and endpoint computer systems. In these attacks, we find three different situations;

- **Distant:** Distant attacks are attacks that come from a location in which the only efforts available are sending information in and awaiting responses through intervening infrastructure. As the attack progresses, more and more proximate and enveloped attacks become feasible for the remote individual, but these are not distant attacks and will not be included here. This includes a wide range of technical attack types including but not limited to:
    - Call forwarding fakes
    - Content-based attacks
    - Data aggregation
    - Distributed coordinated attacks
    - False updates
    - Illegal value insertion
    - Imperfect daemon exploits
    - Induced stress failures
    - Input overflow
    - Network service and protocol attacks
    - Password guessing
    - Reflexive control attacks
    - Viruses
- **Proximate:** Proximate attacks are attacks where the attacker and defender or system under attack are in the same position with respect to observing and affecting information. In this environment, both sides can observe and alter information approximately equally, leading to the ability of the attacker to passively watch what is going on as well as actively alter states or induce behaviors. There are far more proximate attacks than distant ones because of the strong positional situation. This involves a wide range of techniques including but not limited to:

- Cascade failures
- Collaborative misuse
- Covert channel induction
- Data diddling
- Desychronization and time-based attacks
- Error-induced mis-operation
- Excess privilege exploitation
- Infrastructure interference
- Infrastructure observation
- Invalid values on calls
- Multiple error inducement
- Observation in transit
- Privileged program misuse
- Process bypassing
- Replay attacks
- Residual data gathering
- Resource availability manipulation
- Simultaneous access exploitations
- Sympathetic vibration
- Trojan horses
- Undocumented or unknown function exploitation

- **Enveloped:** In the enveloped situation, inputs and outputs are controllable for some pair of parties or for an individual party. In this case different attack mechanisms are available in addition to those available from the proximate position. These include but are not limited to:
    - audit suppression
    - backup theft, corruption, or destruction
    - below-threshold attacks
    - insertion in transit
    - man-in-the-middle attacks
    - modification in transit
    - piggybacking
    - spoofing and masquerading

More information on these attack mechanisms can be found in the security database at all.net, my Web site.

12.3 Perception management

Perception management is about all of the things discussed earlier under propaganda and so forth, only in the case of information attack tactics, it is typically directed at individuals or small groups selected for their utility in attaining the attacker's objectives. In most attacks that most people hear about, there is little obvious mention of the perception issue. But perception management is at the heart of many of them.

- Any time an email containing a virus is read by the users, this is a simplistic perception management attack in which the user is tricked into doing something to aid the attacker.
- Any time someone gets someone to fill out a form over the Internet and that form is used to aide in data collection for that attacker, a perception management attack has taken place.
- Whenever a false email is sent using forged header information and the recipient doesn't notice the difference and acts on the instructions in the email, perception management has taken place.
- Any time a telephone call comes in and the person at the other end of the line gets answers or actions, perception management has been exploited.

Generally, perception management involves an attempt to elicit information, an approach to get someone to act on behalf of the attacker, or an attempt to suppress normal behaviors that would defeat the attacker if exercised. Elicitation is covered in substantial detail in my book "*Frauds, Spies, and Lies*" so I will largely ignore the details here other than to indicate that operations security is used to defeat elicitation and that is also described in the other book. Getting someone to act on behalf of the attacker is typified by the examples given above. Information is provided by the attacker to the target of their attacks and the target is caused to act on behalf of the attacker because they believe that they are doing the right thing through their action.

The other class of perception management actions typically undertaken in these attacks are physical entries, execution of exploitive acts, or placement of exploit capabilities for further use. Examples of typical exploits of this sort include:

- **Be a client:** In this example, you act like a client or potential client and get the royal treatment including tours and similar access that is then used to plant materials or gain additional information.
- **Be a laborer:** Here you fade into the background by looking like other laborers on the site during a construction or repair project and leave exploits or directly launch attacks from there.
- **Be a reporter:** Many enterprises like to tell reporters things and give them the royal treatment so that nicer stories will be written. It helps to have fake press credentials.
- **Be a manager:** Management is always feared and often not questioned by normal employees, so this grants more access, but also risks more attention.
- **Be a supplier:** If you want pricing information, being a supplier and finding out how low you have to go to sell to them gives an indication of the other supplier prices in place.
- **Be a co-worker:** This works even better if you have a badge that looks like theirs, dress like them, and generally appear to be like the other people in their environment.
- **Be a first responder:** Police and fire fighters often get access but they also get a lot of attention, so use this with caution.
- **Come as a group:** It always helps to have someone else who is part of your attack group authenticate you, and targets rarely ask who authenticates them.

These are of course only representative examples of the general principle at work. If it looks like a duck and acts like a duck, it must be a duck. Only it's not.

12.4 Indirect intelligence gathering

Much of the intelligence gathering done on targets is indirect in the sense that it never directly touches the target or provides any indicator of the intelligence effort to them. While governments who believe an individual is an intelligence operative may watch them closely, an operation that is not watched at the source can use a vast array or resources, largely available over the Internet, that give no indication to the target of the intelligence effort. These are indirect intelligence efforts. Here are some of the things we standardly gather from over the Internet in our assessments:

- **Public records searches** are undertaken to identify all of the information about the target associated with any legal filings they have made in any jurisdiction. This includes everything from annual reports for public companies to land holdings to contracts for government agencies, and on and on. For example, on a bid for a contract the bids are released after award so that all parties and anyone else can see all of their contents. This includes lists of workers involved in the bidding process, prices offered, claims made, technologies provided, references, contact names and addresses, and so forth.
- **Credit checks and other similar records** for individuals include such things as taxes paid, income levels, where they live, mortgage amounts on homes and cars, payment history on all credit cards, marriages, divorces, law suits, settlements, salaries and wages, additional income, names of children, schools attended, degrees earned, phone numbers, cell phone usage, numbers called, and durations, restaurants visited and what was bought, stores where purchases were made and what was bought, Internet purchases, Internet sites visited, groups they belong to, places they go, and political donations.
- **Job solicitations and resumes** provide a wealth of information on the company and its technologies and similar features. By looking for Internet job listings that detail technical knowledge and background sought,

resumes of former and current employees looking for other work, and other similar sources, a lot of detailed information can be gathered. For example, computer and network hardware and software details, lists of vendors they work with, projects underway and completed and what those projects were about, names and contact information on other people in the target enterprise, business and technical areas of interest, locations where different activities are being undertaken, how email addresses and telephone numbers work, how people address each other within the organization, and even specific infrastructure projects.

- **Presentations at conferences** are often posted to conference Web sites or in conference booklets. They may include pictures of people, often even wearing their company badges. They may include detailed information on operations, investment strategies, partners, equipment, software, specific configurations, addressing schemes, network topology and firewalls, operations centers, detection and response systems and capabilities, pricing strategies, how electronic data interchange is done, and a wide range of other information of a similar nature. Digital documents in the public domain also often include residual data from prior content of those documents, like previous recipients or contract values.
- **Go a step further** and you can get a lot more information. For example, once you know all of the building locations from Internet searches, find out what building code jurisdiction they are in and look up the public records that provide building plans and layouts. This will allow you to identify data centers, meeting rooms, executive offices, physical security weaknesses, and so forth. Satellite images are also available at resolutions of inches, allowing you to do advanced surveillance and determine approach, entry possibilities, and escape routes.

External indirect intelligence is often done with little or no risk.

12.5 Direct intelligence gathering

The more direct approach is typically undertaken after the indirect approach has been exhausted, unless there is time pressure or the attacker is naive. Direct intelligence gathering involves interactions with the target directly and thus increases the likelihood of detection and response. In information warfare, surprise or having the target never know what happened is generally desirable. Direct intelligence tactics typically involve things like:

- **Interviewing for a job** is an excellent way to learn a lot about a target. With rare exceptions, interview processes lead to detailed conversations with specific experts in the areas of interest, provide limited physical entry into facilities, provide tours, give you opportunities to meet and greet people, involve meals with people that lead to understanding of their habits, and if properly used, they provide details of hierarchies, names of people in positions and what they work on, titles of big projects, approximate release dates, names of vendors and customers, and on and on. Interviews may also lead to jobs that allow you to become an insider at the target.
- **A directory** and hierarchy of the enterprise is often very revealing. For example, the number of people in a business area, their names and phone numbers, and their relationship within the structure lead to a wide array of exploitations involving perception management. The details of people and their backgrounds gleaned from external intelligence leads to identifying the sizes of groups working on projects of interest and their likely funding levels. Phone numbers often reveal locations and proximity leading to additional groupings. People with cell phones tend to be people the company wants to reach in emergencies or that travel a lot. Tidbits like this really add up when forming the mosaic of the target.
- **Checking out business partners** often generates more information on the target than direct contact with the target, but it is somewhat direct and likely to be noticed.

- **Relationship building** is a standard direct elicitation tactic. In this sort of effort, relationships of one form or another are built between individuals and organizations to create larger scale longer-term information exchanges based on increasing trust with time. This often yields insider level access or even management level influence.
- **Becoming a business partner** often provides a very direct level of access to target networks and expertise. It is relatively easy to become a valued partner with any company if you have a front company that is already in business of a similar sort. If I want to gain entry into a network, one of the steps I will take is to become a value added partner through this *front company* and then use the connectivity to attack the target network.
- **Finding pictures** of the people and their names allows many sorts of elicitation methods to be more effective. Augment this with a bit of makeup and a stolen or forged badge and one of your team members can walk into a facility they rarely work at as if you were them.
- **Talking to the landlords** of the facilities they operate is a good idea. In one case I arranged to visit rental space adjacent to a target of interest. You can guess the rest.
- **Line-of-sight surveillance devices** are often usable against targets. This usually involves rental of space.
- **Direct network intelligence** efforts include a wide array of tools and techniques described earlier.
- **Pulling the Web** server and Wayback machine (a Web site) allow the attacker to gather a lot of available present and past intelligence information on a company or location.
- **Dialing for dollars** is an expression used by telephone solicitors, and telephone calls are often used to fish for information.

As a rule of thumb in intelligence gathering, take only pictures, leave only footprints is an understatement. Take all you can get but leave no indicator that you were ever there.

12.6 Garbage collection

I was voted most likely to dumpster dive on my vacation by the College Cyber Defenders at Sandia National Laboratories, where I developed teams that understood these issues in great detail. And in fact, I do indeed dumpster dive on some of my vacations, as part of my consulting business. In fact, as I am writing this section of this book, I am preparing to dumpster dive for a client of mine. Gloves on, boots on, eye protectors in place, and remember to point your toes and hold your nose.

Dumpster diving, going through other peoples' trash for useful information, is one of the most fruitful approaches to attacking information assets. It yields a striking array of materials. Almost anything you can find by complex information attacks has been found in garbage cans and dumpsters of major enterprises, ranging from national security secrets to corporate payroll details.

But paper is not the only thing you find in the garbage. There are plenty of examples of finding active badges, tape and disk media with large volumes of data, access devices, computer systems in need of minor repair, and so forth. These are all potential sources of information that are easy to get.

12.7 Physical entry and appearances

Dumpster diving is at the leading edge of the more general physical access. Access to facilities both at the target and between the target and their collaborators is often feasible and is a highly fruitful avenue of approach for the information attacker.

Physical entry can be forced or facilitated, surreptitious, subversive, or obvious, and aimed at destruction, corruption, denial of services, leakage of content, ongoing access, or use.

Physical entry is often accomplished by appearing to belong and behaving like others in the area. Using influence tactics can be highly successful in gaining entry and support in moving throughout a facility.

The first question you have to ask for physical entry is who you are going to appear to be. The best approach is to figure out who goes where you want to go and what they look like. Then you make yourself look like them and go about your business, ignoring everyone just as they ignore you. If you don't get nervous, you are usually just fine at going wherever you want to go.

Tailgating behind legitimate entries or walking past guard stations often works, as does waiting by an area where workers go outside for coffee breaks and acting like you were out on a coffee break before they arrived, then following them in. This has gotten me into otherwise highly secured (but not secure) facilities.

> Physical entry is the easiest and most effective low tech attack.

As you become recognized, you may be able to return with greater ease, and simply chatting with a guard for a period of time often gets you past security. And of course once one team member is in, they can often open doors for the rest of the team. A clip board often helps to make it look like you belong and causes people to not pursue you because they associate clipboards with people checking on other people.

The great badge scam is one of my favorite methods of physical entry. In locations that use badges religiously, the problem is that once you are in the facility, you need a badge, or you need to look like you have one. The clipboard has saved me many times, strategically held where others wear their badges. But another approach is to find out what a badge looks like and make one of your own to match. Add your picture, their logo, and realistic markings and you have what looks like a legitimate reason to be there. People see the badge and may even introduce themselves, but they will almost never suspect you don't belong in the more public areas if you have a badge.

Finally, there are specific behavioral traits that work in work environments. If you display them you will rarely be challenged.

12.8 Trojan and plants

Once physical or electronic access is gained, it is best exploited for desired goals and to plant ongoing capabilities for reentry. Planting devices in target networks is a typical approach to this. In most organizations, a network device can be placed between any computer and its connection to the internal network with great effect. A properly devised intelligence device will allow remote control from directed energy as well as through covert channels in the infrastructure, have no apparent presence in the location, and only exploit covert channels unless otherwise directed. It should normally be reprogrammable and designed to lose all memory under conditions where it is detected. This is the sort of thing you would expect from an intelligence agency, but not from a rank armature, and detection of such capabilities is a good way to detect the presence of a higher grade threat.

> Information warfare threats are not all quick hit remote operatives working from back rooms. They are often planted over a lifetime to systematically gain access to critical elements of a target.

Physical plants can also be delivered from the factory through factory penetration or interception in transit. A typical computer can readily be modified with hardware or software to provision ongoing Trojan horse capabilities. Either method can be used to emit radio frequency, sonic, or packet-based information to different planted capabilities at different distances. Incoming commands can often also be accepted by such devices. Depending on the specific goals, such a device could be designed to only act when finding material of specific syntax or type and then to collect it over long time frames and deliver it at particular times or under particular circumstances.

People can also be planted within organizations if long time frames are available. Lest you think this is ridiculous, you have to understand that information warfare threats often work in generational time frames. Intelligence operatives have been activated after 20 years or longer in the target country. Consider the information we have described on threats like China, terrorist organizations, military intelligence organizations, and the Central Intelligence Agency. They engage people over periods of their entire careers in these activities.

Some of the alternatives that are commonly used range from building vendor and contractor relationships to becoming career employees. A skilled operator may end up running large portions of the target organization before the end of their career and may do strategic damage to a nation state by running one of its largest industries.

With help along their career, other planted operatives can be leveraged for higher and higher posts in organizations. Someone who makes it to manager can hire in a lot of employees who ultimately get to higher levels in the enterprise, some of whom end up hiring other employees who make it to vice president or even CEO. Group infiltration strategies and tactics applied over long time frames can be incredibly successful over the long term.

In one example I documented, a group of people moved from company to company over a period of about 2 years per company, working their way up to higher and higher levels of responsibility as they moved, taking millions of dollars from each company, infiltrating into the next company, and bringing along other employees. As far as I am aware, they are still operating today and are in their 5th or more company since I discovered them. And this is a small and relatively low funded group of only a dozen folks who produce relatively little real results. Imagine a group where they could produce real results having substantial resources behind them and a national infrastructure of support.

Group infiltration tactics and strategies work in many venues. Over the long run and with adequate resources, they can have national level effects.

12.9 Combinations and sequences

The last example points out that combinations and sequences of methods can be far more powerful than individual methods, and high skills attack organizations such as those found in information warfare apply these techniques to great effect. While most defenders focus on stopping one or two attacks and finding bad guys one at a time, the intelligence problem at the level of information warfare goes orders of magnitude beyond these simplistic approaches, and these threats are very real for all large enterprises today.

> You may only be a target of convenience on their way from here to there.

Criminal gangs and organized crime used the same sorts of tactics but at a less sophisticated level and over shorter time frames of only a few years. They also use violence and obviously illegal methods to gain their objectives. It is fairly common for them to gain a foothold in a business by loaning money at high interest and pressing the collection issue. At some point they gain enough leverage to allow them to take over the companies they loan money to. At that point it is too late for their victims. Leveraging scandals of all sorts and personal destruction, these criminals move into extortion, taking more and more until their target runs dry.

But the skilled information warrior is not after this same objective. They want to gain the same foothold, but they want to leverage it for long term advantage in a very different sort of exchange. They may want to plant Trojan horses in software your company delivers to clients so these can be used to gain leverage into development systems used for military weapons. They may want to gain a better set of positions in other companies and use your infrastructures that connect to those companies to gain additional access to those locations electronically. You may only be a waypoint on the path from their source to their destination. Your infrastructure may only be a way to obscure their efforts to attack someone else.

Combinations and sequences can get rather interesting. For example, suppose a backhoe accidentally severs a cable that has a communications link (it happens all the time) between the primary and backup site. The response might be to move information between sites using a secondary Internet connection through a local Internet service provider (ISP). Now an information warrior might look at this as an opportunity. Break into the ISP, then intentionally create the "accident" that pushes the desired traffic through the infrastructure you have taken over. That's a basic two step plan with a set of sub-plans. Let's look at a 5 step plan:

- Step 1: Plant national level capabilities for disruption of infrastructure by planting people in critical infrastructure jobs.
- Step 2: Work to take over select industries that are key to strategic success.
- Step 3: Seed fear within the country and get political change to happen so that the target expends enormous resources on a minor threat like a sub-state actor.
- Step 4: Use growing economic success to gain financial leverage on the target via loans that support their national debt and trade deficits that create even more debt.
- Step 5: When the time is right, use naturally induced outages in critical infrastructures but exaggerate them with your internal plants, call in the loans and use the economic disruptions to take over their markets, use the taken over industries to leap beyond the target in price and quality of goods, and keep from being directly attacked over these moves by having the target's military fully engaged elsewhere.

That is the sort of thing that strategic warfare is about. As the plans get more complex, they are harder to carry out and more prone to errors. But this can be compensated for with redundancy, momentum, propaganda, and resources. The US knows they are under strategic attack from China, or they should based on declared policies and documents, and yet they are powerless.

13 Legal issues

Many people are of the misimpression that war has no rules, but unless you can really take over the whole world and sustain control of it, you will end up the subject of international law, trade and other sanctions, and other forces that will ally against you. Since nobody has succeeded yet, it is a good bet that information warfare as well as all other warfare will have its legal issues. And so it does.

Generally, each of the actions associated with information warfare may have legal consequences in each jurisdiction involved. This includes the direct criminal and civil implications associated with international law, nation states and unions of various sorts involved in any way, states within nation states and unions, and localities at all levels.

In trying to understand the legal issues, one is faced with a truly daunting task. For example, suppose we want to launch a simple one-step surveillance effort from a private firm in the US against a company in Japan. Here are some of the things that might be involved:

- Trade laws (related to what can be moved to and fro)
- Intellectual property laws (related to what is taken)
- Employment laws (related to worker privacy)
- Frequency spectrum laws (how results are transmitted)
- International laws (a crime according to them)
- Tax laws (did you pay taxes on what you stole)
- Worker safety (do you endanger people when you do it)
- Export laws (is it legal to move across boarders)
- Insurance law (who is liable for what losses)
- Financial records and accounting laws (any requirements)
- Privacy laws (can the records be exported or retained)
- Securities and banking laws (legal releases of data)
- Business records laws (exemptions and exceptions)

All of these laws may differ by jurisdiction and situation and may be invoked by every jurisdiction that any of the effort passed through as well as the source and destination of any particular acts. The US and some other jurisdictions even assert that some laws apply to its citizens in other jurisdictions and even individuals from afar who have never in any way interacted with them or anyone in their jurisdictions. And in many cases the applicable laws are contradictory. All enterprises face a similar problem, but in the higher intensity sorts of conflicts, these tend to get stretched in all manner of ways by the rich and powerful.

Laws also change with time, so tracking all of the changes and their implications can be a heavy burden. It would be nice if there were a resource somewhere that allowed these issues to be settled, but there isn't, and I am not likely to provide one in very much detail here. Nevertheless, I will make an attempt at addressing the specific legal issues associated with information warfare.

But before I do, I should note that most laws are violated by those in power from time to time and generally this happens with little or no recourse. People in power also have a tendency to use their interpretation of the law supported by a set of legal edifices to skirt the laws. A good example of this is the Presidential pardon system that exists in one form or another in most nations. A pardon from the top executive of a country can protect one of its citizens, usually for life, against conviction for any crime, including war crimes and crimes associated with information warfare.

However, there is a caveat. Those who are pardoned are only pardoned subject to the control of the country they are in. Other countries may decide to kidnap them to bring them into a different jurisdiction, when the country falls or political systems change, these pardons may be invalidated, and of course pardons don't prevent attempts at retribution. This is also true of spies who seek refuge, as demonstrated by the Soviet Union, which successfully killed people who committed treason living elsewhere.

13.1 Codes of military conduct

Codes of military conduct limit what military personnel may do in warfare. These codes exist in almost every nation and even in a less well defined form in military organizations of sub-state actors. Even terrorist groups have codes by which they live, typically dictated by a set of religious or other dogmatic texts as interpreted by their leadership, in much the same manner as the nation states have a doctrinal basis and interpretation body for their laws and social mores.

> Psychological treatment of prisoners is one area that is poorly covered by codes of military conduct.

Treatment of prisoners is one of the main areas we hear about in the media when prisoners are not treated according to the Geneva convention. Torture is essentially illegal in all civilized warrior groups, but there is still coercion which is a psychological warfare function. The laws on psychological treatment of prisoners are particularly unclear and of course even in the police systems of most societies, different psychological tactics are used to extract confessions. When asked whether torture is justified in extracting information that could save lives many can be persuaded to adopt torture, but when the psychological community tells us that it doesn't work very well and gets poor information while solidifying the prisoner as a member of the opposition, many consider it to be a foolish tactic. But all agree that psychological tactics like good cop – bad cop and similar stress and relief mechanisms are not only fair but called for in the treatment of prisoners.

Use of weapons of different sorts are also constrained. For example, international law is typically obeyed for those who follow the codes of conduct and this means that various sorts of chemical, biological, and other similar weapons are not permitted. But there is essentially no law or element of the code of conduct that directly addresses information weapons, so in that sense, these weapons are not controlled by the code of military conduct.

Use of deception is controlled. In particular, there are various specific types of deception that are prohibited in warfare. Specifically, the use of false surrender tactics where a flag of truce is raised and the new prisoners use it to draw in the enemy. The use of the Red Cross signal for anything other than a hospital and the targeting of hospitals are also generally considered counter to the rules of war. Military deception is, of course, not only allowed, but strongly promoted, with these rare exceptions. In addition, most military organizations are very careful about the control of deception at a command level because deceptions improperly applied may deceive your own people and deceptions that everybody knows about on your side are likely to be understood by the other side.

Attack on civilians and civilian infrastructure is another far more important issue in modern information warfare. Information infrastructure was, at one time, completely different for military and civilian purposes. They had more or less separate and different networks and communications methods. But in today's world, all of the information infrastructure is fused together in the Internet with some notable and rare exceptions. As a result, the restriction against attacking civilian infrastructure does not apply to information infrastructure any more than it applies to civilian roads used for transporting military payloads.

> The high level of integration between civilian and military information infrastructure makes it all subject to military attack under the military codes of conduct. This means that to the extent that it can be used for military advantage, it will be, and by all sides.

The high level of integration of civilian and military operations, from finance to media control and propaganda, to communications, to energy production and distribution, to networked computers and critical servers, means that all of it is part of the military infrastructure and thus a legitimate target for attack.

13.2 International law and tribunals

International law and tribunals like war crimes trials and decisions made by the World Court and similar bodies are of limited utility in controlling nations. They tend to be far more effective against the losers of wars than against the winners, if only because the winners are rarely available for a trial as they sit in their governments and control the military of their countries.

But international law has different sorts of elements and different sorts of standing in different courts throughout the world, including the court of public opinion that has a great deal of effect on the future of many individuals and nations. International law is currently poorly defined and rules only over countries that agree to it and then only on a case by case basis.

> International law in terms of the World Court, the UN, and other similar bodies is only marginally effective or relevant to information warfare because of its relatively low level of direct control or power.

International law enforcement, in the form of InterPol, has only the ability to coordinate and communicate and has no real power to enforce laws. The United Nations (UN) has peace keeping forces that are comprised of representative fighters from different countries that act under the UN's control. But this control is typically limited to conditions in which the parties more or less agree on the utility of keeping a lid on violence or the desirability of feeding starving people. All of the operational aspects of these UN forces leverage the information infrastructure and intelligence capabilities of the participating nations. The UN is itself a form of information warfare designed to reduce warfare by keeping lines of communication available and providing the aura of legitimacy for actions taken by member states against other states without explicit disapproval by other *leading* states. The presence of the Security Council and it's ability to override any decision by a single vote indicates clearly the extent to which nations have been unwilling to cede power to the UN.

International law also applies to laws between individual countries and laws that countries have with regard to interactions with other nations. And this is an area in which information warfare has enormous implications.

A key example of how international law impacts information warfare is in the area of regulations surrounding cryptography. While countries like France don't allow cryptography within their infrastructure, countries like China allow cryptography only if they have the keys to watch all of the traffic and countries like the US treat cryptographic systems as weapons and prevent export without permits. Cryptography is one of the basic mechanisms used to secure information exchanges. Without it, protection is very hard to achieve over otherwise insecurable infrastructure. Of course this is the very reason that the US limits key sizes on cryptographic systems, while France outlaws them and China demands the keys. They want to be able to have effective offensive information warfare capabilities in terms of reading and inducing signals between communicating parties and they want to be able to defend themselves against the use of information technology by enemies through surveillance of all network traffic.

> Many of the legal restrictions on the use of security mechanisms harm the defenders and help the attackers.

But the problems with these approaches go back to the equities issue. They are very good for attackers who can gain enormous access to information and systems, but they are also very bad for the defenders of those same systems. In the meanwhile, attackers have used steganography to bypass the protection provided by surveillance while breaking into systems by exploiting weak or non-existent cryptography. The defenders, who are trying to follow the laws, are unable to detect or defend themselves against those same attackers, weakening the protections of the critical infrastructures and other information targets that most need to be defended from information warfare attacks. US cryptographic restrictions have weakened US industry while strengthening its enemies.

13.3 Individual rights and privileges

Most people would very much like to think that they have some rights against their own governments, and in the arena of information, those rights typically have something to do with privacy, liberty, freedom to communicate, and similar things. But despite the almost universal acceptance by the citizens of the World surrounding the basic notion of being allowed to do what they want without harming others, the governments of the world disagree.

Governments that many people call oppressive are those that restrict these freedoms and rights. For example, China holds the basic assumption that the government can look at anything at any time and that the citizens have no right of privacy and never have. The European Union (EU) on the other hand has privacy rights embedded in its constitutional structure and culture and demands that no records with personal information be kept any longer than they have to be in order to get the work associated with them done. The US government wants retention of data for investigative and intelligence purposes but business wants to destroy that same data because of the potential liability associated with it being subpoenaed and the cost of tracking it all.

Information warriors are rarely concerned with rights of individuals because most states assume that the rights of the state stand over the rights of the individual, particularly in times of intense conflict. War powers generally override any individual rights and certainly do so for the duration of the hostilities if the leadership in power deems that they are to be ignored. Furthermore, the people punished for such things generally include only the foot soldiers (or in this case keyboarders) and not their leadership who generally ignore these sorts of laws without substantial peril.

Of course the intensity of the conflict and the position of the parties dictates what laws may apply to them in the International arena. And as the saying goes, you have to catch me before you can arrest me. As long as those who violate the laws are not within a jurisdiction that can arrest them they are relatively safe.

13.4 Intellectual property laws and rights

In strategic information warfare over the long term, intellectual property and prowess are perhaps the most important issues at hand. For that reason, intellectual property laws are important to understanding information warfare.

Trade laws are used to limit the spread of intellectual property across boundaries. For example, there may be trade sanctions that prevent the passage of certain classes of intellectual property to specific countries. Some types of intellectual property are not allowed to be exported from some countries. For example, sensitive military technologies are identified by the US International Traffic in Arms Regulations (ITAR). ITAR restricts arms like cryptographic systems but it also restricts things like classified information, any information about nuclear weapons, and a wide array of technologies used in warfare including information warfare,

> Intellectual property issues and rights come into play in a lot of areas of information warfare.

As a completely different example of an intellectual property right that may support information warfare is the freedom of speech. Speech, and its existence as a right is not universal, but where it is considered a right, it is often used to protect the right to seed a country with misinformation or even disinformation, even under the auspices of news. Of course in countries without free speech, disinformation is often used by the government, so this is hardly a differentiator between the different systems of government. What it does show is that propaganda exists and is permitted to exist in all societies. Where laws are needed to protect it, they are in place.

But on the other side of this issue there is always prohibited speech. Even in freedom of speech nations, there are limits to freedoms of speech and people are jailed for such things as sedition or treason for saying things that the government does not want them to say.

False advertising laws are another interesting area. Here is an area where a company that lies can be prosecuted, but a politician who lies cannot. It seems that rights to speech also don't extend to those trying to start riots or those who are leaking secrets, including religious secrets that are copyrighted or trade secrets that are owned by a company.

Free speech on the radio or television will get your license revoked by the Federal Trade Commission if you say any of the 7 dirty words, and if you are broadcasting or even possess images of naked children you can be subjected to laws regarding the publication or possession of child pornography.

> Laws are very hard to follow and don't apply equally to all.

When it comes to things like breaking into corporate computers, there are laws regarding corporate espionage that don't apply to individuals who have their information stolen. But there are laws regarding identity theft that don't affect the theft of a company's identity, even though corporations are considered persons in the US for most legal purposes. Privacy laws, on the other hand, apply largely to personal identifying information about individual human persons and not to information about corporations which have private information, but not about health related matters and only apply to financial information until the annual report.

Generally there is a lack of precedence for legal issues associated with most of these laws in information warfare circumstances, but there is a great deal of precedence for many of them in criminal activities. This brings us to the notion that most of the things identified in this book as part of information warfare may be illegal or legal depending on the specific circumstances of the actions. Financial records laws and frequency spectrum laws are other examples where the range of information warfare activities go from legal under almost all circumstances to areas where the information warfare methods or materials are illegal. For example, it is illegal in the US and Canada to possess a van Eck machine.

13.5 Patent, copyright, and trade secret laws

Intellectual property laws typically cover patents, trademarks, trade secrets, and copyrights. Each exists for a different purpose:

- **Patents:** A patent covers a process or a means and method and is designed to assure that the owner of the patent can control its use through licensing while advancing society by notifying all of the new development it brings. The patent holder gets control for about 17 years while the rest of the world can apply the patent to many fields of use. A patent, when submitted, has to be novel, the best way to do something, and enabling in that it allows those normally skilled in the art to implement it. Patents have to be published to work.
- **Trade secrets:** A trade secret is something that is so clever and yet unknown to anyone else, that it provides business advantage to the holder through its secrecy. Once disclosed, all protection is lost, except for suits against those who release the Information. Trade secrets must be controlled by proper marking, control over access, and similar methods in order to be enforceable.
- **Copyrights:** Copyright laws provide for control over duplication of any specific writing, depiction, sonic recording, or other similar realization of anything that is reduced to tangible form. For example, this book is copyrighted. Copyright protection starts whenever something is put into tangible form and lasts something like 99 years past the death of the author.
- **Trademarks:** A trademark is a marking that is used to identify a particular item or enterprise. For example, a phrase, symbol, picture, shape, or other object can be a trademark once registered, and this provides protection against its unauthorized use. When a large company like Coca Cola spends hundreds of millions of dollars

promoting its name and the shape of its bottle around the globe, it protects its investment with the trademark.

Together, these four forms of intellectual property protection form the set of available options for assuring that someone who creates value by turning their ideas into realizations is protected from others taking advantage of their efforts to gain from their labors without renumeration. They exist, ostensibly, to provide for the welfare of all by assuring that those who put forth intellectual effort are rewarded through ownership, just as those who put forth physical labor to build something own the resulting property. Thus governments provide reward guarantees for advancing society.

Governments typically exempt themselves from many of these laws, but not always. For example, a US company sued the CIA for selling a modified copy of its software (with an implanted Trojan Horse) to the Australian government for their taxation work. The company won. Governments also favor their own industries in many cases, for example, China does not enforce the smart card patents internally, and they make billions of smart cards for use all over the country without paying royalties.

Patents are hard and expensive to get and to enforce. As a result, they are largely used by large enterprises to create barriers to entry from other large enterprises. They are also leveraged by small patent holders to extract monies from large enterprises when the patent holder often didn't deserve the patent in the first place. But in rare cases, patents are used by legitimate small inventors to gain well-deserved recognition and royalties for the work products of their minds.

On the other side of the spectrum, many companies use minor patent changes to extent patents for additional years, copyrights are hard to enforce and are often violated in huge volume over the Internet, and are sometimes over-leveraged to prevent fair use, which is the copyright provision that allows the use of copyrighted material for scholarly work, for research, and for other uses – like quotations in newspapers. Clearly, intellectual property law is a complex area in which many participants seek to leverage legal systems and agreements for competitive advantage.

13.6 Wealth and power and the law

As the intensity of conflicts wane, many more things are considered from a legal standpoint than at high intensity levels. The law is often less harsh for greater harm in times of greater peril. For example, when at peace, issues of intellectual property come to the fore, while at time of war, theft of intellectual property not associated with the war effort is generally less attended to. As the level of intensity lowers, more and more things become less and less clear in terms of the law. In fact, almost everything identified in this book that is illegal in peace time may be legal if performed by a military organization during a war.

And many things that are illegal for one group may be legal for another. For example, the codes of military conduct apply to the military and not to contractors that work for them. Insider trading laws apply to those in corporations but not those in Congress. So pending legislation that is about to pass with a minor but critical alteration at 4 AM is not subject to insider trading sanctions when the members of Congress or their staff invest in the company aided by the legislation before anybody else has a chance to read the details. As the price goes up from the legally mandated advantage, the congressional staff pulls out before the second piece of legislation that disadvantages the same company shows up. Never confuse laws with justice.

In every nation on Earth since the beginning of time, wealth and power have allowed individuals to reduce or eliminate legal implications of their actions while those without the wealth and power to prevail have been labeled as criminals and sent to jail for the same or lesser acts. In the information warfare arena, this is no different than anywhere else. White collar criminals who steal many millions of dollars often end up in jail for only 3-5 years while blue collar criminals who steal only $500 end up in jail for 15 years and in some cases, defendants who only scanned a computer for known vulnerabilities have been sent to jail for many years while those who regularly scan those same systems for profit in corporations are rewarded with news stories and never prosecuted.

14 Information warfare defenses

I know that I have given inadequate attention to defenses against information warfare attacks along the way through this book, and I didn't want to leave them out entirely. So the remainder of this book is dedicated almost entirely to defenses and how the people of the world can protect themselves from the horrors of information warfare.

> There are many defensive methods and approaches and they come in a wide range of areas and types.

Defenses are particularly problematic for a number of reasons, not the least of which is that the uncommonality of objectives means that by defending one side it may actually help the other side. Defense takes resources, so if one side can cause the other to expend resources on defense, they will gain advantages in efficiency by not having to defend themselves. This means that a good defense will also have an offense that forces the opponents to defend themselves so as to not lose the advantage of inefficiency to the other party. So the successful strategic defense will force the opponents to also defend by including an attack component strong enough to force the other parties to defend as well.

Many of the mechanisms of offense are also viable mechanisms for defense. For example, censorship is used to suppress undesired ideas as part of propaganda, but counter propaganda also uses the same technique to reduce the amount of information provided by the propagandist.

Defenses can be strategic or tactical, long term or short term, preventive, detective and reactive, or adaptive, they can address life cycle issues with businesses, systems, people, or content, they can be in the form of business protections, psychological protections, political mechanisms, or any of a wide range of other things. Defenses, like attacks, are strategies that combine many methods together in a coordinated protective effort.

14.1 Technical defenses

Technical defenses constitute an enormous range of methods designed to deter, prevent, detect, and react to attack and to adapt over time to improve those defenses. The database of defenses at all.net includes 140 classes of techniques. For presentation, I will divide technical defenses into four different categories; structure, perception, content, and behavior. These four categories represent only one way of looking at these issues. Generally speaking, these defenses are the mechanisms that come into direct contact with the content or the mechanisms that store, process, or communicate it.

> Creating and operating a set of technical defenses requires a serious effort over a long time and involves a lot of specialized expertise and resources commensurate with the risks being addressed.

Many of these defenses are cross-cutting so that they have effects in more than one of the identified areas and beyond the direct contact with content. A lot of controls are directed at people and processes. People controls are discussed throughout this book while process controls are used to assure that systematic and repeatable methods are used to increase the chances of things operating as designed. Many of these techniques can be found in the all.net database, and I will only quickly review some of them here.

At the governance level, policy development, the creation of control standards, compliance with laws and regulations, business continuity and disaster recovery planning, risk acceptance, transfer, avoidance, the integration of multifaceted defenses, the creation and operation of internal controls, fusion of multiple disciplines into a cohesive approach, the timeliness of detection and response, a tracking process for evaluating performance, and a desire to keep things simple all help to build a meaningful program. This governance is necessary in order for protection programs to be effective, and it is best covered in another book of mine titled "*The CISO ToolKit – Governance Guidebook*"

14.2 Technical structural defenses

These defenses generally include mandatory and discretionary access and flow controls, firewalls, and other barrier mechanisms. They are generally associated with the separation of one thing from another so that they don't interact or so that they interact only in well defined places and ways. This is an appealing approach because there is good reason to believe that separation prevents causes in one area from producing effects in the other. If we can find the proper way to separate things, they will be protected from each other. The techniques include:

- Authorization limitation is used to limit what an authenticated party is authorized to do.
- Automated protection checkers and setters detect and report on deviations from authorization policy and correct them.
- Chinese walls are used to separate functions that must not link.
- Classifying information as to sensitivity is used to bundle information with different properties together for handling in bulk.
- Controlling physical access is used to enforce separation.
- Disconnection of maintenance access prevents its exploitation.
- Drop boxes and processors are used to securely hold content.
- Effective mandatory access control enforces logical separation.
- Faraday boxes prevent electromagnetic information leakage.
- Fault isolation limits the effects of faults to a locality.
- Fine-grained access control allows detailed control over access.
- Fire doors, fire walls, and asbestos suits prevent fire damage.
- Increased or enhanced perimeters increase attack difficulty.
- Independent computer and tool use by auditors prevents internal exploitations from going undetected.
- Independent control of audit information prevents authorized users from corrupting audit trails.
- Information flow controls limit where content can go.
- Isolated sub-file-system areas restrict content within file systems.
- Limited sharing restricts what can be shared with whom.

- Limited transitivity prevents giving away received content.
- Lockouts prevent risky actions during maintenance periods.
- Locks are used to prevent access to areas to those without keys.
- Minimizing traffic in work areas prevents exposure to threats.
- Minimizing copies of sensitive information reduced the chances for leakage or damage.
- Multi-person controls limit possibly harmful acts by individuals.
- Multi-version programming prevents single faults from causing systemic failures
- Path diversity provides redundancy to compensate for faults.
- Periods processing and color changes prevent mixing of content that must be separated.
- Physical switches or shields on equipment limits harm to that equipment from outside sources.
- Placing equipment and supplies out of harms way limits the sources of failure.
- Secure or trusted channels provide assurance that communicating parties are who each thinks the other is.
- Suppression of incomplete, erroneous, or obsolete data prevents its reuse or replay when not appropriate.
- Separation of duties limits the effects of individuals.
- Separation of equipment limits damage from localized events.
- Separation of function limits the functional impacts of faults in any given component.
- Tempest protection prevents wave forms from going where they do not belong.
- Temporary blindness separates systems from each other during periods when trust cannot be reliably established.
- Trunk access restriction limits the exploitation of communications trunks.
- Trusted system technologies provide separation mechanisms with defined levels of surety.
- Waste data destruction provides coverage for the end of the life cycle for content.

These defenses exemplify the range and utility of separation mechanisms for limiting the effects of attacks and accidents on content and its utility.

14.3 Technical perception defenses

These defenses focus on how content, systems, situations, people, and things are viewed by the different people and systems viewing them. They generally involve understanding how people and systems view of their environment leads to their behaviors and controlling these views so as to control the behaviors. These defenses include but are not limited to:

- Accountability creates the impression that what is done is accounted for and things undone or overdone will be found and attributed to the responsible party.
- Awareness of implications provides the means by which individuals can understand the personal and non-personal implications of their actions.
- Clear lines of responsibility for protection provide the ability to identify who should do what so that the promise of punishment can be fulfilled.
- Concealed services prevent potential exploiters from determining that exploitable functions are present.
- Deceptions are used in a wide range of ways to induce or suppress signals so that the attacker becomes ineffective.
- Document and information control procedures provide clarity as to who has what so that others can see when someone else does something inappropriate.
- Effective protection mind-set provides the awareness and understanding necessary to allow people to act to protect the interest of the enterprise.
- Feeding false information is a deceptive method that causes others to consume resources wastefully.
- Improved morality increased the likelihood that people will not act in ways that are knowingly harmful to others.
- Individual accountability for all assets and actions links the individual directly to their actions, placing a guarantee that they

are aware of that when they act inappropriately, they will get caught.

- Infrastructure-wide digging hotlines provide information on where not to dig to avoid breaking communications lines.
- Jamming creates the impression that signals are not present or are unaccessible if present.
- Legal agreements provide formal notice of obligations and intent to carry out those obligations with explicit remedied for failure to meet obligations.
- Low building profile reduced the interest in specific facilities and makes their import less obvious to potential attackers.
- Noise injection changes the signal to noise ratio so that signals appear to be not present or are hard to identify and gather.
- Numbering and tracking of all sensitive information provides a clear and obvious means for identifying when something is missing, who last had it, and where it is supposed to be.
- Protection of names of resources makes it more difficult to identify what is what and its import or meaning.
- Retaining confidentiality of security status limits the ability of the attacker to determine what may or may not work in what circumstance and therefore what attack mechanisms to apply in which circumstances.
- Security marking and/or labeling provides clearly readable and obvious identification of the sensitivity of content and through association with badges and location, whether the holder should have the content in the particular place.
- Spread spectrum is used to spread the signal over a broader electromagnetic spectrum, thus concealing it within a broader range of wave forms.
- Training and awareness provide the perception of what individuals should do and the ability for them to identify when individuals are doing things they should not be doing.
- Universal use of badges provides identification and marking that associates individuals with access and belonging.

Together, these and other similar methods provide effects on the perception of attackers and defenders that produce behavioral characteristics more likely to result in effective protection.

14.4 Technical content defenses

These defenses address the meaningful utility of the material being sent, stored, or used. The nature of this challenge implies that such defenses will be imperfect because only content without utility has only one legitimate outcome. If there are more legitimate outcomes, then there is no way to tell which of them is correct for the situation, the utility of the content lies in its differentiation between legitimate options. Content defenses include but are not limited to:

- Change management limits the ability to make changes to approved sets of individuals who go through appropriate processes to verify that those changes are appropriate to the need.
- Authenticated information is used to increase the certainty with which the content can be determined to be as assumed.
- Authentication of packets provides low-level authentication of the source and content of packets of Information containing useful content.
- Configuration management is used to identify inappropriate configurations according to technical security policies and to correct those configurations to the appropriate settings.
- Content checking provides independent verification that content is as it is supposed to be.
- Encrypted authentication provides hard to forge verification that content is authentic as to source and not modified in transit or storage.
- Encryption provides concealment of content from those unable to decrypt the content without a proper key and those with proper keys but not in possession of the proper decryption capabilities.
- Filtering devices are used to remove undesired content from information flows.

- Inspection of incoming and outgoing materials provides assurance that those materials are free from hazards and are suitable and appropriate to their movement.
- Integrity checking provides verification as to source, authenticity, propriety in context, non-modification, and reflection of reality.
- Integrity shells are real-time integrity checking mechanisms used to detect alteration between verification and use.
- Known-attack scanning detects known attack mechanisms before they can cause further harm.
- Out-of-range detection detected variations outside of expected values for the context so that they can be investigated further before being trusted.
- Protection of data used in system testing provides for independence of tests and limits the ability of attackers to make alterations that pass tests even though they are inappropriate. It is also used to limit the potential for leakage of content.
- Searches and inspections are used to periodically or upon identification of due cause, do in-depth verifications of the propriety of content.

These sets of defenses provide particular attention to the useful content to assure its utility. None of them are or can be perfect in the sense of assuring that all systems always do the proper things. Rather, they provide a defined level of certainty associated with specific properties of that content within a select context. Unlike structural defenses that have a physics or similar hard scientific basis, content and perception defenses are based on less certain facets and properties of information and are more directly tied to context.

14.5 Technical behavioral defenses

These defenses work by seeking to understand and differentiate legitimate from illegitimate behaviors. They deal with people and systems and are typically designed to detect and react to events or as overriding controls over behaviors.

- Anomaly detection seeks to detect things that just don't look right according to the normal behavioral patterns of the environment.
- Alarms provide announcements of detected events defined as relevant to the defender.
- Auditing provides reviews of behaviors of systems and individuals to detect deviations from identified legitimate activities.
- Conservative resource allocation allows behaviors of resources to be better predicted and avoids deadlocks and most resources starvation failure modes.
- Detection before failure uses an indications and warnings methodology to identify indicators of failures before they occur and warn of the impending failures in time to mitigate the resulting harm.
- Detection of waste examination is used to determine when someone is trying to use waste products to gain content or intelligence information.
- Disabling unsafe features limits the available features that can be exploited so as to limit the behaviors of the system.
- Least privilege is a sort of behavioral constraint that limits the capabilities or privileges of an individual and the processes acting on their behalf to the minimum privileges required for them to do their work.
- Limited function applies special purpose devices instead of general purpose ones to perform specific functions, thus limiting the potential for exploitations to those functions designed into the mechanisms.
- Misuse detection seeks to identify and report unauthorized or inappropriately applied uses.

- Over-damped protocols are protocols that automatically reduce the quantity of content on each subsequent round of exchange so as to prevent resource exhaustion and expanding loops.
- Properly prioritized resource usage applies mechanisms to assure that more urgent and important things have priority over less urgent and important things.
- Quotas are used to limit the consumption of resources by individuals and groups.
- Redundancy provides a means by which behaviors can be assured even in the presence of failed components.
- Rerouting of attacks is used to prevent attempts at interference from causing interference by handling the attacks in a different part of the infrastructure.
- Secure distribution provides a means by which content can be distributed with increased certainty of arrival in tact and on time.
- Strong change control limits the mechanisms of change so that inappropriate change is harder to undertake.
- Testing provides independent verification that content and systems meet properties defined for them.
- Time or use variant augmented authentication provides increased certainty of authenticity as consequences increase.
- Time, location, function, etc. access limitations limit who, what, where, when, why, and how content can be used.
- Traps temporarily limit activities to localities under tight control.
- Trusted applications provide higher levels of certainty regarding specific properties of their operation.
- Trusted repair teams provide sets of people that are trusted to perform specific repair and maintenance functions.
- Uninterruptible power supplies provide assurance against momentary outages in power or disruption of the normal wave forms of power.

Behavioral defenses range broadly but in the extreme, reach the least certainty of any technical defenses available. As such, the more extreme behavioral defenses are very soft, and yet they also provide the sorts of uncertainties for the attacker that make them interesting and useful.

14.6 Psychological defenses

The notion of psychological defense is a particularly interesting one in that it is inherent in most conflict situations but it is almost never brought out as a separate subject. But in the arena of information warfare, psychological defense is clearly a vital and fundamental component. Psychological attack is at the core of propaganda and politics, and every war ever staged started and ended with psychological operations.

> We have no choice. If we are going to live in the information age we must learn to cope with the psychological issues that it brings.

How then do we counter the massive efforts exerted to control what we think about and what we think about it. While it could be said that this is an age old problem, it has been greatly exacerbated by the recent changes in Western civilization, specifically, the emergence of the information age. Today it is impossible for most people to earn a living or carry out their daily lives without interactions with information technologies and thereby the media. You cannot go to the bank without getting an advertisement, you cannot use electronic mail without spam, you cannot make a sensible vote in an election without wading through the hyperbole, and you cannot drive to work (for those of you who do) without passing hundreds of signs. There is almost nothing in our society or lifestyle that is not used to influence us and manipulated to favor one or another interest.

I remember when I was young. I used to walk ten miles barefooted through the snow on broken glass up hill to and from school, but we never saw any advertising on our walks! OK – I admit it – I had shoes... My point is that the times have always been changing. Bob Dylan has a great song that says things like "*you'd better start swimming or you'll sink like a stone, the times they are a changin*'". The times are indeed changing, but they have always changed and they always will. People that claim that the pace of change has accelerated probably weren't alive during the industrial revolution or the French revolution or the plague years.

Psychological defenses involve a dramatic change in the way the world is understood by those living in it. Clearly those who are ignorant of the influences being exerted against them will be influenced and will have little power over their lives. And those who learn the ways of the farce, those who come to understand how to exert themselves over the minds of others will prosper and come to power and influence, just as they always have. So this then is the mission of the elite psychological information warriors. To become so skilled at their craft that they can induce desired behaviors in others on a global scale. And it doesn't matter whether you are on the offense or the defense, in psychological operations, these goals are the same. The battle for the will of the people, the hearts and minds war, uses the same weapons for offense as for defense.

But what do the rest of us do? As a normal person, I have only two possibilities. I can either become cowed by the peddlers of influence or become one of them. You are with them or against them, by their own declaration. While some may be able to educate themselves and become passively defensive by ignoring much of the media and spending a lot of time and effort searching for real information and others may seek to avoid the battle by acting like it doesn't exist, It impacts their lives in every way, and they are in fact cowed by it. Consider that the laws they live under, the available products on the market, the services they can buy or sell, the places they can go, the things they can see, and every other aspect of their lives is essentially ruled by the perceptions of others and the persuasive efforts of the influencers. You can live reasonably well for a while under these pressures before they start to squeeze you. You can choose to hear less, do less, and suffer under the yoke of the oppressors, but they will engage you eventually whether you want to be engaged or not. They will take your children and force them into wars by psychological operations against them. They will corrupt them with promises of bliss, whether in the role of suicide bomber or defender of freedom, they are all just as dead and just as taken by the persuasion police.

Induce fear and propose a solution, that is the way of influence. And that is what I am doing by writing this book. It's not that I want you to be afraid, it's that I want you to change. The world is the world. By keeping the truths of it from you, those in power retain their power, and I want to seed a revolution in your mind. I want the revolution to change your mind about how you deal with the world you live in. Viva la revolution! Long live changing your mind. The change I want to make in your mind is to have you struggle against the mental oppression of those who seek to influence you, and eventually, I even want you to rebel against my views.

> I don't want you to change your mind about the things you have been thinking about. I want you to start thinking about different things.

As I described earlier, there are two critical facets to mind control. The first is to control the subject being considered and the second is to control the considerations about it. Most of those who are actively and intentionally fighting the information war that is eternally underway in human society seek to control how the subjects at hand are portrayed and considered. But this is itself highly deceptive in that it limits your thinking and limits the consideration to the things they want considered. The revolution I want to seed is far more seditious to their controls. I want you to change what you are thinking about, not what you think about it.

The goal of my revolution is to get you to listen to what people are talking about briefly, and then start them talking about something else, something important to you, but not something they have been talking about. If a politician tries to engage you in a town hall meeting or some other discussion and brings up a subject, I want you to engage in the discussion by bringing up a completely different issue that you care about and expressing that what they are discussing is less important to you and most of the people you know than [place your favorite subject here]. I want the stale debates over divisive issues to end and to be replaced by thoughtful discussion of topics that matter to your life.

Now it turns out that my strategy is not just for politicians. It is also for propagandists (who are also usually politicians, but that's a different matter). My revolution of the mind extends to propagandists as well. When they assert their psychological methods on you and others in a group, I want you to revolt by challenging them directly on the psychology they are using. Don't just stand by and passively let them influence others. Tell everyone else what they are doing, the mechanisms they are using, and how they are using these techniques to put forth their lies and propaganda. Change the subject from whatever they were discussing to their use of propaganda. Change the subject to the erroneous assumptions they are making. Change the subject to their motives for misleading the people who are present. Ask them who is paying them to do what they are doing, and know the answer in advance along with who is really funding their front company. Put the facts in front of everyone to expose them.

How about the news media? Is it good enough to turn them off? No, it is not. Because when you turn them off, you are leaving the rest of us to suffer. The Christian coalition does it better than most. They do group protests against the sponsors of news and programs they don't like. Here's what I want you to do. I want you to revolt against the media like the religious right of the US does.

In refusing to be corralled by the persuasion police, it is also very helpful to recognize and counter their tactics. Again, this is not hard to do once you recognize what's going on. If they try to divide and conquer by using issues that break people apart, unify the group by bringing up issues that most people agree on and imply or state that the persuader is trying to divide and conquer. If they try to marginalize you by being aggressive and tagging you with something, counter by telling everyone present that their attempts to paint you in an unfavorable light by lying about you only reflects how disingenuous they are, that you have as much right to express your views as they do, and that you resent the sort of abusive tactics that they are using to marginalize you. We have little choice but to engage in psychological warfare. It's think or be thunked.

14.7 Business defenses

Businesses can realistically only defend themselves. It is not the task of business to defend a nation or a people or anyone other than its shareholders against loss of their investments. And that means that businesses must optimize their use of defenses through a process of risk management.

To most readers this will sound harsh, but I see no other way to portray it. It's not that businesses can't have and apply the principle of enlightened self-interest. This notion is generally associated with the idea that it is good for a business to do things that ultimately come back to help it, even if they may seem like short term negatives. For example, when a business donates to a charity, it is quite literally a give-away, but it almost always comes with publicity and a public relations campaign that is designed to improve the public image of the company and therefore to bring benefits in terms of customer loyalty and increased exposure to its customer base. It may also lead to meetings with others who fund these charities and these relationships may lead to more business.

Business cannot ignore competitive issues and survive.

But when it comes to issues like who will manufacture electronic parts and what they will cost, businesses that fail to address the reduced cost of Chinese parts compared to their internal costs will ultimately fail. My brother-in-law went to a manufacturing firm in Ohio that makes custom parts. They were outstanding at what they did and there was no doubt of their expertise. Their price was excellent and far below the competition in the area. But they came out on the first day and said that as soon as the quantities reached a few hundred, the cost would be lower by an order of magnitude if the parts were outsourced to China. If my brother-in-law can sell them for $50 and his cost is $25, as soon as the business gets large enough to warrant competition, someone will have them made in China for $2.50 and sell them for $24 and my brother-in-law will be out of business.

In the information protection arena, business has to have a risk management process in order to be effective at self defense. All businesses have to have adequate self defense because there are real threats, real vulnerabilities, and real consequences to information attacks. The typical business defense process should follow the process of a standard like *COSO*, which stands for *The "Committee of Sponsoring Organizations" of the Treadway Commission*. This committee is a global group that developed and published an *Enterprise Risk Management Framework* that was ultimately identified as the archetype by the regulators implementing the Sarbanes-Oxley Act in the US. It is as close to a high level standard for how business should address risk as exists today.

> Information warfare defense for business is a governance function fulfilled by internal control processes.

Risk management is a critical guiding principle and process that any business has to undertake to make sound decisions. All top level executives have some level of expertise in risk management and those that are better at it end up more successful in business because those that fail to address risks well end up losing more often and they don't rise to the top. Of course it could just be luck and connections that brings people to lead enterprises, but when this happens the enterprise pays the price and those individuals go or the enterprise fails. Something about survival of the fittest where fitness in risk management is defined by survival.

Businesses use governance processes to define the utility of content to their business, associate duties to protect that utility, use risk management to determine what to protect and how well, and then use enterprise security management as a control system to produce reasonable and prudent risk management, and thereby sets of defenses where appropriate. The individuals tasked with protection then use power and influence to fulfill their duties to protect and generate feedback to assure that those duties are fulfilled by those responsible.

14.8 Military defenses

Military organizations are tasked with different responsibilities in different countries. Typically, military organizations have:

- A supply and logistics element that has to be protected and this includes all of the content and technology that supports that content, including interdependencies.
- Most countries have a limit on military control over civilian assets such as:
  - The means of production of weapons
  - The information that supports operations
  - Power and heat
  - Water supplies
  - Financial infrastructure
  - Manufacturing infrastructure
  - Clothing and housing
  - Food supplies
  - Civil government
- A need to protect tactical information from disruption corruption, and disclosure for fairly short time frames and more strategic elements of planning and weapons systems capabilities, designs, and operations for longer time frames.
- A dependency on command and control structures that have to be protected in order to continue functioning.
- Intelligence capabilities that provide them with the key information they need to be able to make sound decisions.

Successful attacks on any of these things can cause military defeat even for the strongest military forces. And every military organization in the world knows this. For these things to go right military organizations must defend them successfully, and their opponents will seek with reckless abandon to disrupt these elements of military power in order to win the conflict. In order to defend these things they need the funding over long time frames to create and operate existing capabilities and develop new ones.

If there is one thing to understand about military operations, it is that innovation and technical advancement have always and will continue to make the difference between winning and losing. All military organizations have enormous strategic logistics tails in the area of technical advancements of all sorts and the dire need to keep these advancements controlled so that they come to the advantage of the side developing them. It would be naive in the extreme to imagine that this will somehow change and that the world is reaching a new level of peaceful coexistence where this is no longer the case.

A strong research and development program producing meaningful scientific progress that is applicable to military use in the long term must be augmented with an effective information protection program in order for strategic success to be achieved. By going back and reviewing the earlier discussion on how research and development has been going and understanding the increasingly interdependent world being created by globalization, it quickly becomes clear that the next serious global military conflict that lasts for more than a short period of time will result in dramatic changes in world leadership.

Military defenses end up dealing with the most serious sorts of threats and those threats will attack interdependencies with violence or intelligence as necessary in order to disrupt military operations. But it is impossible to defend at all points all the time. So it is necessary for military organizations to use intelligence assets to understand more specific capabilities and intents of enemies in order to successfully determine what to defend against what mechanisms and how well. This active defense combines with broad-based defenses at a minimum standard level in order to provide a capacity to deter, prevent, and detect and react to attacks, so as to retain military advantage. All of the defensive techniques discussed earlier apply but when and where they need to be applied depends on what is important to defend and how well, in much the same way as it is for business. Risk management is undertaken but the threats and consequences are far different.

14.9 Modeling, simulation, and gaming

One of the things that military organizations have come to understand with time is the criticality of modeling, simulation, and gaming for understanding the strategic issues, both on offense and on defense. On the defensive side, modeling is used for a wide variety of things. A good example is modeling for understanding the interdependencies and criticality of these interdependencies. The same modeling methods used to understand how to attack an enemy can and should be used to understand how an enemy can attack you. This then becomes a critical tool in planning defenses.

Modeling and simulation will increasingly become key elements of defensive information warfare at a strategic and national level as nation states come to better understand these issues.

Simulation plays a variety of roles in defensive information warfare, but it is less used than simulations for offense, largely because offense is generally given a priority over defense in warfare because the enemy cannot kill you if they are already dead. But when killing the enemy is not the goal, military organizations are not as good at getting the job done. And intelligence processes and other related aspects of the governmental, military, industrial triad start to become more important to success.

One area where defenses require simulation is in the area of detection and response processes associated with information attack. Another area is in chemical and biological attack where response processes are key to survival of large populations of cities and the ability to sever transportation is critical to the success against longer-term biological threats. The response of the US to September 11, 2001 was actually an outstanding example of slowing transportation so as to give the defense time to react, and it worked very well. You can expect simulations of these sorts of activities and dry runs will become more common around the world in the information warfare arena.

14.10 Indications and warnings

As we come to understand the situation through models and simulation and gaming out the scenarios that may arise, the ultimate objective may be to move from a system of detection and response to a system of indications and warnings.

Detection and response is generally focused on waiting to see effects of attacks and then coming up with response processes that mitigate the harm and reduce the length of the incident. But the obvious problem with this approach is that it always waits for an attack to take place and only then responds to it. For high speed attack this can be devastating and of course you are guaranteed to lose something along the way during the time between when the attack starts and the time before detection and response are effective.

The idea of indications and warnings is to get ahead of the curve so to speak, by understanding what is indicative of a coming attack and providing warning to those who have to deal with it before the attack arrives. In other words, the detection happens before the attack arrives and the response occurs so that the attack never succeeds.

Today, in the information warfare arena, we have plenty of indicators of attack. The US and India would have to be blind to not understand that China is attacking them. Any nation state that doesn't understand the US capacity for information warfare and its ongoing commitment to it must be asleep at the switch. If governments don't see the information attacks of terrorist groups that are coming on a daily basis, it can only be through their own inability to open their eyes. If politicians don't know in advance that they will be under information attack before they start a campaign, they are not suitable for their jobs. As this book points out, the war is underway, and we know just what sorts of things are coming, from where, how they work, and how they will be used. And I hope that, if nothing else, this book provides strategic warning to all that if they fail to act in defense they will suffer the consequences.

14.11 Political defenses

Political defense is easy in a dictatorship or similar hierarchical society. All differing ideas are crushed and internal assets are directed at spying on citizens in order to detect any deviation. The media is exploited to the maximum to create points of view supporting the leadership, and all sorts of mock elections, projections of opulence, claims of victory, and other similar propaganda methods are applied without limit. From the womb to the tomb, you are part of the state and the state is part of you. The notion of freedoms, if they exist at all, are projected as the ones that the politicians want you to see that you have, and other freedoms are identified with their negative impacts. Information is tightly controlled, to the point of creating special purpose mechanisms to allow Internet access to the citizens while blocking portions of the Internet that are undesirable. If people try to encrypt or obscure information to allow exchanges of undesirable ideas, they are watched, collected, re-educated, or otherwise put out of business quickly and harshly. Social stigmas are attached to dissent, and success is tied to supporting the powers that be.

> Now if you don't like how the US has turned out, you can always leave and try for a different sort of political society. Just kidding... it isn't that bad here yet. But give it time.

In forms of government that support more freedom of expression, political defenses against information operations focus on dominating the discussion and limiting it to the issues that the politicians can compete on and discuss effectively.

Few politicians are masters of more than a few topics, so they need to keep the discussion on those topics in order to have the aura of infallibility and vast knowledge. If the media started to ask questions about serious global issues like starvation, disease, the various wars underway in parts of the world your politicians are ignoring, the long-term issues of clean water availability, the looming shortages in oil and other sources of power, the global changes associated with pollution and climate issues, lost jobs, and so forth, it would make for a very complex political situation indeed.

And all of these issues are only the start of the real issues facing the nations of the world on a tactical scale. Much larger issues are looming relating to the long term place of science in human society, religious fanaticism, educational directions, the exploration of space and of the oceans, protection of rain forests and biological diversity, the lack of crop diversity and the need for protections against the upcoming environmental changes, changes in land mass available for supporting food production, rising ocean levels, changes in the electromagnetic field of Earth, the effects of increased travel on the animal and plant population of Earth, genetic controls over humans that will soon be available and how they should be limited or applied, psychological impacts of family and lifestyle changes on future generations, how to care for the aged, the dramatic increase in asthma cases, how population increases associated with longer life spans will be handled, the encroachment of humans on the last vestiges of wildlife areas and its impact, changes in the oceans and how it impacts the microscopic processes we depend on for survival, defense against comet strikes, the increasing rates of skin cancers, sound pollution, political corruption and bribery, the increase in crime and criminal enterprises and how to deal with them, border protections and the movement of masses of populations associated with local changes in agriculture and finances, breakthroughs in physics that could lead to far more destructive accidents, robotics and the future of human society with increased intelligence in computer systems, and I hate to keep on going but we are nowhere near running out of issues.

In the increasing world of knowledge and its ready availability to so many people all over the globe, it may be to the point where none of these things are going to be, strictly speaking, controllable any more. Political defense nowadays largely consists to trying to limit the discussion and focus the attention of people on the things you can meaningfully claim to address even though they are not addressed at all. And it increasingly surrounds generating more money and appeasing those with money so you can stay in power to facilitate the rich getting richer. Where have I heard this before?

14.12 Individual defenses

Individuals can attempt to defend themselves against information warfare attacks, but their capacity to do so is severely limited. As a first approximation, individuals have essentially no control over their own information because it resides in other hands. When an attack on the credit card company occurs, my information is leaked, and I have absolutely no control over any of it. The only option I have is to use cash for everything and this has largely been made illegal or very hard by many nation states that require certain transactions to happen electronically, that require direct deposit into bank accounts, that mandate electronic filing of certain information, and that limit carrying cash across borders.

> We can each act to protect ourselves in simple ways, like shredding documents and not posting personal information, but little can really be done against information warfare.

If I want to have access to any modern convenience, I need to pay service providers, such as telephone, gas, electric, water, garbage collection, and other similar companies. I need to have a bank account to do this in many cases, however sometimes I can pay in cash for local public utilities. While some have selected the cash only approach, even they can no longer travel by air or train, medical care is problematic, they cannot get prescription drugs, and cannot buy or legally operate a car without a driver's license. Even the Amish have to have licenses on their horse carts, and the Luddites still have to pay taxes. All of us are codified in one way or another in databases and these databases affect our lives directly and indirectly. And we have little control over any of these.

The only control that regular people have comes in the form of elections of public officials who may make laws that alter the structure of things over a long time frame, but that control is extremely indirect and tenuous because essentially all of the candidates are supported by and beholden to other people and big businesses. Political systems have always been corrupt by their very nature.

14.13 Media defenses

Another area where defenses may be feasible is in protecting the media and its independence and freedoms. The free press is a critical component of information warfare defense in that it provides the means for independent information gathering, analysis, and presentation to become available to the populace and decision makers.

> If reporters were sincere about getting the truth out, they would stop taking stories they are fed and find their own stories. When 60 minutes producers decided not to air a major issue, the whole team would walk off and take the news to the other media outlets.

The media is supposed to protect the freedom of speech by exerting it, and the courts are supposed to support this effort in a free country, or so the old saw goes. But the media is dominated by a small number of corporate interests that limit discussions of issues that effect them. For example, when the Tobacco industry was purposely adding addictive substances to cigarettes to increase the physical dependency of its customers and was denying it in legally binding statements, the news magazine *60 Minutes* had a story from an insider that would have blown the scandal wide open. But the corporate types decided that they didn't want to air it because they figured they would be sued by the Tobacco companies. Now this is a news magazine show that has never shied away from such controversies when it comes to issues that don't have close links to their corporate governance committee members, but big Tobacco was too powerful for CBS to go up against.

The large number of media outlets does mean that the news is out there to be found. I get news feeds from the Internet from all over the world, and the things going on around the world are vast indeed compared to the things discussed in the evening news and other highly successful commercial media outlets that dominate the viewing public. It takes money to attract the audiences and this means you need money to make money. And the people with money protect their own interests.

14.14 National and strategic defenses

As I think I have now described defenses to a reasonable extent, the defenses of enterprises, military organizations, and individuals, political defenses, media defenses, technical defenses, and psychological defenses. It should be clear from my descriptions and views that I believe that these defenses do not individually address national or strategic needs. In other words, the strategic information warfare defense of the nation state and its population against the high grade threats are not covered by the individual defenses or by combinations of these defenses through some sort of serendipity. The solution to national strategic information warfare defense must therefore reside at the national level, and that means that a cohesive strategy created, supported, and executed by the overall national government is necessary in order to be successful in information warfare.

> Nations that thrive and survive have strong strategic long-term defensive information warfare capabilities. Those that don't have and use these capabilities suffer and ultimately degrade or collapse.

National governments that fail to create and operate a long-term strategic as well as tactical defense in information warfare are destined to fail. This is a pretty stark statement to make, but it is not hyperbole or exaggeration and I am not likely to take it back. In fact, I will go a step further and indicate that the nation states that are successful at information warfare are highly likely to also be successful at coming to dominance over time and that those that fail in this arena are almost certain to fall into disarray and reduced national standards of living and global influence. In other words, up to a fairly high level, I believe that there is a strong correlation between success of a nation state and the prosperity of its people and its ability to do an effective job of strategic defensive information warfare.

> If I am right, this means that strategic national defense against information warfare is an imperative and that failure to act in this defense means the destruction of your nation.

I don't mean to say that with this alone a nation state will prosper. I do mean to say that countries with only limited military capacity have managed to survive and thrive while states with enormous military might have collapsed and that these successes and failures were largely due to the intellectual capacity of those nations and their ability to retain those strategic information assets and defend them against other nation states. Another way of saying this is to say that intellectual capital is the single most important element in success for nation states and undefended, this capital will be overtaken by offensive information warfare efforts springing from both within and without.

From the issues I have been bringing up in this book, you would think that there is a war all around us and that I am asking you to fight a battle on all fronts and all at once. You think right. There is a constant war surrounding all of us today and it is the war over our thought processes, our intellectual property and capability, our business processes, our strategic investments, our information and information systems, our infrastructures, and the deeper intellectual interests of our societies and our people.

There is a crying need for understanding these issues by everyone who will be a part of society in the information age. The future success of nations and the well being of our children and their children lies in our ability to understand these issues at this point in time and act on them at a strategic level.

So what do we do about it, and where does the strategic future of nations lie? I believe that it lies in two absolutely critical areas and that any nation state or people that fails to appreciate and emphasize this above other issues will ultimately be unable to sustain itself. These areas are education and science, and that is what the rest of this book is about.

15 Education and the future of the World

There can be no doubt that the rise of humans to the top of the food chain represents a clear demonstration that intelligence and the ability to apply it effectively is more important to survival than speed, numbers, or brute force. And even those who believe that humans were implanted instantaneously on Earth by God stipulate by their dogma that humans are in charge because of their superior knowledge and ability to manipulate the world – in the image of an all seeing, all knowing, all powerful God.

> Those that ignore the lessons of history are destined to relive the failures of the past.

As in prehistoric times, humans learn and apply what they learn through an educational process that brings the young up to the knowledge level of their parents and, in select areas, well beyond it. That is the way humans survive and thrive. And as human civilization moved forward and humans came to dominate the Earth, the need for specialization also became clear. No person on Earth, even 5,000 years ago, knew enough to do everything that humans everywhere could do. And as the pace of innovation increased and technical areas arose, this specialization became more pronounced and the need for additional education increased with it. Students assisted masters in their field for periods of years before they became good enough to do the work and become a master on their own.

In the emergence from the Middle Ages, the variety of science and technology for buildings, agriculture, breeding of animals, castles, cities, water systems, lighting, housing, policing, military sciences, mathematics, chemistry, writing, history, medicine, librarianship, and other related areas flourished and required increasing levels of education in order to sustain and grow. This ultimately led to public and private educational systems starting at the elementary level, moving to secondary education, and several hundred years ago into the realm of the University. More and more of peoples' lives were being consumed in education prior to their useful working lives. And as progress was made, people started to live longer, leading to more success for the better educated.

And of course the weapons of war progressed along with the rest of society. The survival of group after group depended increasingly on warrior technology and the educated members of the warrior groups. Those that think Ghengis Kahn was a wild man from the woods who overcame higher levels of technology to win at war missed a major part of the history of these conflicts. Lightning strikes with horse-bound experts bearing specialized weapons used speed and tempo, superior weaponry and maneuver, and better battlefield tactics to defeat their enemies. Castle technologies progressed dramatically over time to meet the increasing mobility of forces and increasing capabilities of siege engines. In each of the stone, bronze, and iron ages, weapons technology advances led to military victory and failure to keep up led to defeat.

The industrial age produced yet another revolution in education and science as people came to realize that increasing capacity to make thoughtful decisions produced a more efficient and a better and more productive industrial base. Industrial age wars also produced dramatic demonstrations of the utility of education and science on victory at war and the industrial bases of nation states along with extreme specialization produced military organizations that were irresistible by those with less education and science. World War 2 produced ultimate victory only because of the vast advances of science by the allies to meet the advances in science brought about by the Axis powers.

In war as in peace, success and prosperity are closely tied to advancements in science and these are brought about by investments and advances in education. The society that creates and maintains an educated populace wins and the society that forsakes education and science collapses. This is the lesson of history and those who fail to appreciate it are destined to suffer the consequences of their ignorance.

15.1 The need for an educated populace

In the information age, as in every age of the past, the education of the populace is the fundamental determinant of future success. The best military in the world will not hold the advantage for long if it starts to fall behind in the strategic hearts and minds war of an educated populace.

But the information age does bring some differences that are particularly worthy of note. One of the keys to understanding is that in the information age, the very heart of the advantage comes directly from knowledge and the ability to apply it, and not from the indirect application to technology. In other words, information is not only the key to success, it is a weapon and a target. Our educational systems are the weapons of information war and they are also the targets of information warfare.

> Ignorance leads to despair, desperation leads to hate, hatred leads to violence, violence leads to collapse of the social fabric, and the collapse of the social fabric leads to ignorance.

For those who wish to defeat a nation, there is nothing more powerful than a movement toward religious dogma and ignorance that destroys science education. It drives leading nations into a spiral of loss, frustration, and economic despair that eventually leaves so little prosperity and little motive to do more than survive that nation states sink into terrorism and radical religious leaders are able to take seed as the only remaining hope for the plight of the poor and disenfranchised.

The only way out of this spiral to failure is education, but it can be applied in a positive or a negative fashion. Germany created one of the best educational systems in the World in its prelude to World War 2. The people of Germany were educated in hate and the sciences necessary to support the war effort until they became so full of technology, hatred, and dominance that they were able to bring the world to war. The German educational system had dogma dominating science and history and the result was terrible war.

Germany is not the only example of a strong educational system ruled by dogma instead of scientific inquiry gone wildly awry. Another one of the best recent examples is the Soviet Union. In the Soviet Union, the educational system was outstanding in that it brought top flight experts to bear in doing science and education, but it failed in another way. It determined at an early age which people should be educated and in what areas. The dominance of the State and the anti-religious fervor of the system combined with the dogmatic adherence to belief in their society and supposed *Social* or *Commune* 'ism', another pseudo religious view, led to their collapse as their controlled governmental economy combined with the prolonged cold war and attempts to keep knowledge from the people.

> The sure signs of collapse seem to start with the inundation of the educational system with religious or pseudo-religious dogma, a reduction of or attempt to control science education by non-scientists, and the unification of education around a central set of standards for performance and content.

Meanwhile, we also have excellent recent examples of successes. Some of the best of these examples are India and China. Both have created strong educational systems dominated by science education, and essentially controlled by their scientific communities with strong support from their governments and social pressures to become more educated in the areas of science and technology. These strategic efforts over periods of decades have resulted in them leapfrogging over the science and technology base of their major global competitors. When combined with their low labor rates, the result has been a dramatic shift in national wealth, prosperity, and global power and influence. And these countries are only still in the beginnings of these efforts.

Over the next 20 years, these programs will reach another generation of their populations and over the next 40 years, there will be more highly educated Chinese and Indians than there are educated people in the entire rest of the world. They may well have already won the Global strategic educational battle.

15.2 The need for education in specific areas

Education, like other areas, can be considered in terms of niche advantages. Additional education in China on US culture might be very important to their information warfare advantage, while US Education on Indian religions may be of little value to its advantage. The areas of education depend heavily on context and on information warfare strategies. Nothing can likely be ruled out as having some import, but on the other hand, some things are clearly of import to information warfare regardless of the specifics of the conflicts.

The difference between education and training is also vital to understanding. When I discuss education, I am talking about things you learn that stay with you throughout your life and career, as opposed to the specific details of today, which are really only exemplars of things to come. While it could be argued that technology changes so fast that today's education doesn't apply for very long, this represents a lack of clarity on education and its value. My education in the 1970s is extremely relevant today in the 2000s and things that were discovered and invented a thousands years ago are still relevant today as they will be for a long time. Newton's understanding of physics, while imperfect because it fails to consider relativistic effects, is still applicable to almost all of the design and implementation of physical devices today. Mathematics from hundreds or thousands of years ago is still applicable as well. All of the understanding I have of finite state machines from my university experience applies today as it did then. My psychology courses, as limited as they were, still apply to modern issues in propaganda, and the history of the world has not changed very much in the last 50 years. This is the nature of a good education.

There are also things I learned in my educational career that are not of much use today. For example, in the assembly programming language section of my comparative computer languages course, the PDP-8 assembler code I wrote at the time has little utility any more. But the experience of writing those programs prepared me for deeper understandings of issues that remain vital today.

I have also had occasional training courses. For example, I was trained at one point in the use of a particular handgun. I don't remember any of it, largely because I learned to shoot long before this and the details of this gun are not used enough in my life to warrant my brain keeping them around. I was also trained how to tie knots in the Boy Scouts. But I don't remember much about it even though I got my merit badge and could tie a lot of knots. The things that stuck with me were not the details of the knots but the experiences I had in the scouts, like the experience that taught me that chopping wood warms you twice, which I will remember till the day I die.

The strategic question that nations need to ask themselves is what education they should give to their citizens to help them win the information war underway now and likely to continue for a long time to come. Will it be a deeper sense of religious values or more computer science? Is it psychology that is vital to the national interest or physical education? Do we need more expertise in manufacturing or automotive maintenance? What do we want to support and what do we want to pay for?

Along with this notion of what to support comes a set of issues related to research directions. The most successful universities historically have mixed advanced research with education. This leads to a faculty team that is highly skilled in research and has substantial practical experience brought to them by external consulting or development aspects of their research. They combine this with a strong set of educational requirements that bring this knowledge to the next generation.

Education in specific areas leads to strategic advantage in those areas, and success in the information age mandates dominance in information technologies, but this is only the start. Information technology is closely linked to all areas of science. And education is not only about the 5-10 year time frame. Education leads to strategic advantage over the 20-50 year time frame. Remember that it takes 20 years or more and lasts for another 60 years.

15.3 The need for an educational system

This is truly a stunning differentiator between education and other areas of endeavor. Even building infrastructures like bridges, tunnels, and highways are typically only 2-10 year efforts designed to last for 30-50 years. Education at the Bachelors level takes 20 years or so, and you need to add another 2 for a Masters degree and another 4 for a Ph.D., an M.D., or a J.D. And that assumes you are going full time. The return on investment only starts to come then, and it takes another 10 years or so before it beats out the total value returned for a typical student who graduated from high school and got a job as a car mechanic or apprenticed as a plumber.

The reason you need an educational system and not just some education here and there is that every nation state is in competition with other nation states over the strategic education of its citizens. In the US, I was educated under the strategic policies of post World War 2 leadership. The result of their efforts is reflected in the citizens now in their 50s and above. The people who earned Ph.D.s in my era are now the senior level professors in US educational institutions. They are teaching the most advanced courses to the most advanced students we have. They are, quite literally, the seed corn for creating the future of the US in its quest for success in information warfare. So why is the US doing so poorly compared to the Chinese and Indians in the information warfare arena?

When I got my doctorate, I was the only US citizen to get a Ph.D. in an information protection subject not including cryptography for two years before and two years after. I personally represent 4 years of the total US educational system output in the information protection arena at the doctoral level. Now start to look across the board at people who graduate in different fields and the progress 20-40 years later and you will see that the investments in US education in information warfare-related subjects compared to that of other countries has been so poor that the US is at a large disadvantage and will likely remain so for the next 20 years.

World War 3: We're losing it...

The US is ill prepared for information warfare, but the world is competitive, so in order to put this in perspective, we need to look at output from other countries. While I was in school, there were a lot of others in the same schools I went to in similar subject areas. Most of them were Chinese, Indian, and Europeans. They got degrees in the same subjects and went back to their countries where they became part of their national capability and helped to grow their national future in information technology, engineering, and other related subjects.

Their educational systems were built up by the same effect at thousands of educational institutions across the US and elsewhere in the world. And over the period of the last 30 years the countries that invested in educational systems, not just individual educators or programs, but wholesale educational systems for their people, are prospering today. China and India have more and better experts in most aspects of information technology than the US does today because the US failed to invest and the Chinese and Indians decided to invest. They spent the resources required for strategic information warfare advantage and the US did not, and the US is now just starting to pay the price.

I am now at the age where professors get their full professorships and have the most important contributions to their societies. And so are those Indian and Chinese students who went to school with me. They are making their greatest contributions and doing so in larger numbers. And this includes helping to create more experts and passing their knowledge along. The US cut education for the generation after mine and undermined it in mine in this area, and they left you and me and our children with a serious deficit that we have to make up for in a very serious way. And we have to do it soon or we will lose the expertise that we have. That is why we need an educational system, and not just some education here and there. But don't look for quick solutions or expect them. The people we start to educate today will be at my level in 2055 or so. The US has a lot of catching up to do. To do it, the US needs a long-term, systematic, and stable approach to education.

15.4 The need for a professorate

Education at the advanced level only succeeds if there is an adequate body of professors available to provide the knowledge, research, and guidance required to succeed. The notion that we can solve the educational problems with automation or technological advances is an interesting one, but it has proven elusive as more and more people have tried and failed to do education over the Internet.

> If universities started requiring and enforcing intellectual property rights, they would be huge financial institutions that would dominate science and technology.

The truly sad thing that I see is that people who are the leaders of today in the US, all of whom were educated by dedicated, under-paid professors who had graduate degrees in their fields of specialty, have little or no respect for the educational system and even less for the educators within it. At all levels, teacher pay is less than the pay for most of their graduates, and the *academic* view is often scorned by industry. But the academics have taught almost every industry leader and most of their employees for most of their young lives. If academic institutions were to require patent and copyright contracts with their students and enforce intellectual property rights on all of the achievements made by their students and material taught in their classes, they might well own much of the world. But they belong to a different community, one that largely believes in sharing knowledge for the benefit of humankind. Of course these communities survive because they benefit the greater societies they serve, and those societies dictate whether the Universities prosper or not by their strategic approaches.

Universities are run by the professors. Those that have tried other approaches have failed because they cannot obtain and retain high quality experts in their fields with doctorates unless they give them the respect they are due. They end up with mediocrity in their professors and their graduates.

But the professorate needs to change and adapt to the world they live in. If they become monks like the religious separatists before them, they will lose touch with reality and lead only meaningless institutions. And of course most professors do change. They tend to stay on top of their fields by attending workshops, seminars, and conferences all over the world and interacting on line and more directly with their colleagues around the world. They live in a system of peer review where other professors from all over the world with expertise in their fields review the work they send to publication, and they have to generate several major publications per year or perish as junior professors. It typically takes 5-7 years to get tenure, a situation in which, by the time it arrives, sees an assistant professor in their 30s being promoted to associate professor with a promise of a job for life. The purpose of the job for life is to assure that the professor has the freedom to pursue their field without outside peer pressure forcing them into stale lines of research.

> This precious resource is the key component of a successful strategic information warfare capability They are the seed corn we need to survive.

Academic freedom comes at a price of course. Peers in the academic community are there every day, in meetings, in classes, in conferences, and so forth, and their reviews sting for the truly radical. Pay raises are not automatic and the professor who more or less goes to sleep after tenure arrives finds themselves teaching large undergraduate classes in service courses with enormous teaching loads and low pay for the rest of their career. For those who have the desire, there is promotion to full professor from associate. A full professor is near the top of the food chain in the University, has higher pay, more respect, and even more freedom to do as they wish. They still have a teaching load and salary depends on performance, but by the time they become a full professor most are about 50 years old. They have been teaching and researching successfully for more than half of their lives, they have published hundreds of refereed papers, they have generated at least tens of millions of dollars of research funding, and they have taught thousands of students.

15.5 The need for an educated industrial base

The reason we desperately need the professorate is because they are the seed corn that grows the future professors and educates our workforce. Our workforce, if properly educated, produces ever more useful work products. The nation state that fails to produce an intellectual class and fails to bring their intelligence into the workplace will ultimately be unable to make anything as well or efficiently as the competition in other countries. Because transportation is so relatively low cost in large volume, this means that the undereducated workforce can only succeed at making small runs of custom and semi-custom goods and performing services for other people. This leads to importing more than you export which leads to trade deficits that leads to financial destruction if it goes too far. Of course inflation follows, reduction in the value of currency, and global leadership falls away to becoming a second world then third world nation. This is the fall of a society.

> The rise and fall of civilizations now follows 30-50 years behind the rise and fall of their educational systems.

But societies rise and fall with time, or at least it has always been so. Societies that fall may never rise again, but all societies that rise, arose from somewhere lower. The people of China and India get paid less than the rest of the industrialized world and their lifestyle is poorer for now. That means that they are highly motivated to work to improve their lives, and given the opportunity to become educated, parents send their children to learn how to improve their lives. These educated students become the key members of the next generation of their industrial base and produce the innovations that gain their people better lives. They invest in the educational system to generate more and more educated people. The more people they educate, the better their industrial base becomes and the better they are able to compete with the rest of the world. They rise as a society and become the competition for the societies that once led the world.

15.6 The need for science and engineering

In understanding what specific areas of education are vital to a nation state and therefore form the strategic necessity, there are clearly many points of view. This book is about information warfare, but I don't mean to demean all of the fields I don't mention along the way. They are perhaps less critical to information warfare in the direct sense, but perhaps just as or more important in the sense of having a civil society with appreciation for the things they win in their conflicts.

> There is no field of science or engineering that is not vital to information warfare, and without a strong science and engineering educational system, nation states are destined to succumb to enemy attack.

Clearly, science is a critical area where progress in one area leads to progress in others. Information warfare is greatly informed by every field of science, ranging from biology to zoology. The so-called hard sciences must be understood and exploited for attack and defense in effective information warfare operations. Electronics and electrical engineering, civil engineering, architecture, chemical engineering, computer science (which is a misnomer), and all of the related areas are all vital to information dominance. Fields as diverse as mining and underwater operations become critical success factors in information operations, and the ability to exploit knowledge in any of these fields either for offense or defense has proven vital in case after case. The ultimate information warfare may be biological and genetics warfare. In these extreme views, the information that is at the heart of all life forms becomes subject to attack and defense. Genetic engineering (a joke on a General Electric commercial) - it brings good things to life! I'm certain I am giving short shrift to many other fields by mentioning these, but clearly if there is an area of science or engineering, it can be used in an information attack and must be understood in order to carry out successful defense.

15.7 The need for mathematics education

Mathematics is under-respected and hated by many who have problems with doing the complex manipulations and perhaps who are also ticked off at how it is taught. I am among them. But I have had something like 15 years of mathematical education, and I survived it and can even apply a fair amount of it when I have to. But whether you or I can do mathematics with the best in the world is not the issue. The issue is that mathematics is central to the ability to make sound decisions in any field of endeavor and is at the heart of technology innovation, engineering, and science as well as analysis required in order to deal with psychological experiments, the ability to understand statistics, and the ability to effectively and efficiently manage. It is critical to simulation that is necessary to modeling and analysis of possible futures, it is vital to the measurement of efforts, and if any country falls significantly behind in its ability to apply mathematics, it is guaranteed to fall behind in other areas rapidly.

There are also great shortcuts for those who want to apply mathematics but perhaps don't enjoy spending hours solving complex but well understood equations. One of the areas where computers have done a magnificent job of replacing human expertise is in the area of the standard sorts of mathematics used in most engineering and science. There are programs available for free that solve the mathematical equations we often encounter in these areas in a very short time and with detailed descriptions of all the steps they made in deriving the solutions. An example is the *Maxima* software program. Other commercialized versions also exist for more specialized solutions in particular fields. And automation exists for much of engineering mathematics and analysis. What we need in great numbers is not people who can wade through the equations, but rather people who know how to set up the mathematical models for situations and apply the existing programs to solving the existing problems. In other words, you need the Education, but you don't need to spend the rest of your life on the symbol manipulations. You have to understand it in order to apply it.

15.8 Soft science is the hard science

I spent much of my career thinking that the so-called hard sciences are harder and more important than the so-called soft sciences. But I now firmly believe that the soft sciences are harder to do than the hard sciences and that education in the soft sciences is absolutely critical to success in information warfare.

> There is a lot of research in psychology that needs to be understood by more people in order to free them from the bonds of psychological slavery.

The psychological aspects of information warfare are undeniable. The hearts and minds war goes to the heart of what information warfare is about and is the only realistic way to use information warfare methods to reduce or eliminate conflict or to sustain the support and technical advancements required to win the global information war. The first battle to wage in the strategic information conflicts of today and tomorrow is the battle for the future of the educational system. If this battle is lost, then the war is lost. And this battle will not be won by logic and reasoning because the people that have to be convinced do not respond well to purely logical or rational arguments about the long term future of society or humanity. They simply don't care about it. They will, of course deny this, because they do care about appearing to care, about everything that's important to their supporting voters.

The psychology of persuasion is the real battleground for information warfare among human beings and persuasion is increasingly being understood in society after society through the efforts of soft science. These efforts are, of course, predominantly carried out by university researchers. The results are applied most by commercial companies in their advertising and media campaigns. And the battle for the future of research and education in the area of persuasion is on. The battle is on for our hearts and minds, and if you and I don't come to understand the persuasion efforts being used upon us, we will all end up the slaves of psychological influencer – the cowed public serving the master.

15.9 Secret science – the oxymoron

The ability for free people to remain free hinges on the elimination of secret science. I will explain soon, but before I do, I should answer the obvious question. Why does this belong within the section on education? I know that by asking the question, you have already guessed much of my answer, and that it is bad persuasion to answer before I ask the question or to fail to create a need then fulfill it. But you know by now that creating a need and fulfilling it is one of the things that persuaders use to get you to think their way. And this issue is important to me, so I don't want to use any trickery to convey it to you. If I did, you might see my trickery as insidious and decide to ignore what I am trying to tell you. Of course this reverse psychology may be seen as a trick as well. How can you tell?

> The research and educational system of institutions of higher learning provide a means by which science can be rewarded without hobbling it.

Open science is practiced in Universities and in most other educational research settings. It works so well because the underlying assumptions, bases for refutation experiments, and experimental techniques and results are all published in a timely fashion. These results can then be confirmed or refuted by independent experiment in order to gain scientific validity and increase understanding. This should also help to answer the question of why University researchers have such a problem with patenting all of their results. Science progresses more slowly when it becomes a method for keeping other people out or taking money from them. And science fails when it is done in a closed forum because, while an individual may prosper, science as a whole and humanity as a whole prospers less and more slowly. So open science is better for all of humanity. But there is a big problem here in that the person who does the science is not rewarded by such a system, unless the system finds ways to reward them for their scientific efforts. And that is what a University or other higher educational system does that business cannot do.

But there is, as always, another snag. When you sponsor Universities to do research and development, it puts the non-University research and development team at a disadvantage. They don't have the same infrastructure and government funding to support their efforts and the University may put out something for free that they worked for years to develop for commercial use. I have had this happen to me. Back in the 1980s I put out a product that used cryptographic checksums to detect corruptions of content in computer systems. As it was starting to become successful, a major university announced a free product that, while it was not quite as good, was free, and had the support of the University public relations machine and of the researcher in on line forums. I was ultimately beaten out of the market, and a few years later, after becoming widely established, the product from the University went commercial. I was screwed.

Intellectual property law is supposed to protect people from things like this, but patents are quite expensive and take a long time to become useful unless you are large enough to enforce them. The key here is not that the University did anything wrong. The key is that some sort of reconciliation between the need for published, peer reviewed scientific research and the commercialization of ideas. If Universities are going to be in commercial industry, they need to be for profit, pay taxes, control their intellectual property, and call themselves businesses. If not, the society has to recognize that the University has a place as a non-profit institution, plays a vital role, and gets adequately funded to prosper and to bring its prosperity to our society in a fair manner. That means funding for research and education, but separation from commercial enterprises.

Secret science is not science at all, because it cannot be confirmed or refuted by independent experiment. Public science is necessary for a credible legal system to use scientific evidence, for success in information warfare, and for progress of a nation state relative to the rest of the world. But public science means public funding, and this means a national strategy for scientific research and education.

Secret science must exist today because governments have a need for advantage over other governments based on some sort of information advantage. Asymmetric advantage is the thing that prevents sheer numbers from winning wars and secret science provides a good deal of that advantage. While secret science has its place, secrets are far too often used only to cover up incompetency and prevent independent review.

The place that secret science has been used of late in the US is in the courts, where witnesses for the government claim that secrecy trumps the independent judgment of the courts and jurists and the rules of evidence do not apply. The government presents a conclusion claiming that there is a scientific basis that cannot be revealed and asks the courts to accept the secret science as if it were open science subject to peer review. The Daubert challenge, which says that expert witnesses testifying to scientific Information may only introduce evidence if the scientific community has peer reviewed the basis of the science and it is widely accepted, holds up in most courts today, as it should.

Secret science cannot be challenged and forms a sort of religion. I have seen this in my work in classified areas of organizations. No matter what the good idea is, someone will ask how you know it hasn't already been done in a secret project somewhere else. Of course nobody can know things that are other peoples' secrets, but most experts in their field have a pretty good idea of the limits of their science and the likelihood of such secrets existing in them. Not invented here means it does not exist. Secret science means we can always claim to be superior. Classified information means I might know more than you do. It's all a bunch of baloney. I have worked on highly classified efforts of various kinds and, with extremely rare and highly niche oriented exceptions, secret science is no better than open science, and it is often far behind. When the classified briefings I attend summarize the open scientific work I did years before as if the world was unaware of it, it shows just how vacuous the secret sciences are. If only for their own good, secret science should almost always let the bright light of day in.

15.10 The last place to look for funds

When countries start to run into debt, have less income than outlays, and are holding a war, they start to cut all spending on other areas. And because there is no real lobby for education in most countries other than through things like unions, the political interests have a tendency to outweigh the public and national interests, even in times of high intensity conflict.

> *"The nation that stops funding long-term education and scientific research in order to support a war, either already lost the war, or is about to."*

Sun Tzu didn't write that, but he might have. Actually, I just wrote it. It reflects the situation in the US today, and it is a sad commentary on a society when it decides that the future of its children are less important than immediate needs of a military conflict.

When countries start to cut education and limit their government funded research and development to only items that support the war effort, it means they are desperate. England did it in World War 2, and it was clearly justified. So did many other countries. Desperate times call for desperate measures, and it is generally better to save your children from starvation and death than to have them better educated – in the short run. But in the long run, our children are better fed and live longer and prosper better when they are better educated.

In the Information warfare arena, cutting education, limiting research to areas supporting a war effort, and other similar signs of desperation that show short-term goals overriding long-term needs of the country are the best indicators of a country on the verge of collapse. Whether the US is on the verge of collapse or not, I cannot say, and I certainly hope it is not, but if the current moves in the US government are rational, then this is the only valid conclusion that can be drawn. Of course if the premise is false, that's another issue.

16 The end is near

At least the end of this book is near...

Many who discuss limited information warfare issues tell their listeners that the end is near, so naturally I had to match their rhetoric with a chapter of the same name, otherwise I would not get as much media attention required in order to pay for the cost of manufacturing the first 50 copies of this book. Of course I don't think that the end is near, I think that the information war is eternal. It is an eternal struggle for supremacy where you and I try to succeed in peaceful coexistence despite all attempts of the World around us to try to put us at conflict with each other.

> In information warfare, all you and I can really do is think globally and strategically and act locally and tactically.

I plan to live quite a few more years and I certainly hope that you do. My view is that information war is trans-generational, so I want my children and their children and your children and their children to live a great long time as well, and to do so in freedom of expression, thought, and with an eye toward future generations well beyond their lives. I fully expect that there will be conflict for as long as people continue to be imperfect, and that is certainly a long time from now unless the Earth gets hit by a really big comet or some similar thing before humanity goes to other stars and planets.

While I would certainly choose to live at peace with everyone else and in eternal scientific and personal discovery and ecologically balanced harmony with the Universe, I don't think the human world is yet ready to allow me to do it. So the only alternatives I have are to be crushed by the wheel of information warfare or to struggle to resolve the conflicts while keeping my family and legacy as safe as I can. So how can you and I do this for ourselves and the World we live in? The answer seems to me to lie in the notion of thinking globally and strategically, but acting locally and tactically.

16.1 The information warfare A-team

In information warfare, team approaches are key to success. Nobody knows everything and nobody can effectively do everything that needs to be done to protect even a small enterprise. This book is well over 300 pages and is only an introduction to information warfare. An encyclopedic volume just covering these subjects in reasonable depth would likely be several hundred times this length, or in the tens of thousands of pages. While I hope someone at the NSF will call me up tomorrow and offer funding for the next ten years to create that encyclopedia, I doubt that it will happen. And if it did, who would read it?

Success in information warfare demands a team approach. Groups of individuals with the right collections of skills and the ability to work well together can collaborate to change the world. In limited attacks, very small teams can be highly effective at going after points of interest, and often single individuals gain capabilities like great influence over many politicians, as Jack Abramoff admitted in court, or control over hundreds of thousands of computers all over the world, like Jeanson James Ancheta, a 20-year-old Californian, admitted to in court. These large-scale information attacks undertaken by individuals can have enormous impact, but they are not likely to turn a war or have the sort of strategic impact necessary to settle a global information conflict. To do this, it takes a larger and more effective team.

If the goal is changing the hearts and minds of billions of people, the influences necessary to do this must be global and must be coordinated. Such an effort must be trans-generational because change happens slowly in the hearts and minds war, and stable peace and discovery can only happen when eduction gains enough of a foothold that it almost becomes a religious doctrine in and of itself. The religion of non-religion is not what I had in mind.

Rather, the world of the future will be led by groups of people around the globe who, overcoming their governments and institutions, manage to form loose global alliances and cooperate.

16.2 Thinking globally and strategically

In order to be effective in Information warfare, it is necessary to think beyond the realm of today and the realm of your personal experience. The notion of thinking outside the box is not what I had in mind here, but rather, thinking in a bigger box. We are all human and constrained by our cognitive limitations. But we can and must join forces in groups of collaborating people and machines to form structures and systems that allow us to work increasingly complex and larger scale longer term issues together in order to address out futures together.

Thinking globally and strategically means that a group must be formed across the world. Like the current terrorist groups and organized criminals of the world, the independent information warriors of the world must unite for the future of humanity if they are to be successful. And because they will, more often than not, be facing powerful political and criminal forces that array against them for personal gain and advantage, the independent information warriors must create their society like so many of the other insurgencies of the world, in secret and with cut outs, so that if any one of them gets caught, the others will remain relatively safe. By using the technologies of the Internet and their skill and knowledge, they must join forces in the greatest covert collection of non-criminal minds ever to populate the planet. A civil organization of independent leaders in their fields who will shape the future of the world across generation after generation. Here is the part where I exude an insane laugh and press the button that will lead us to Armageddon.

But I am a peaceful person, don't wish to run and hide from the governments of the world for the rest of my life, and realize that the only real hope is to start the process of discussing these issues in workable ways. The solution must somehow come from influencing the world, and thus I write. That is the end result of my strategic thinking for now. The pen is mightier than the sword, and the only way to work the magic of influence is by inspiring others. That is the strategy for a peaceful future for humanity.

16.3 Acting locally and tactically

Having said this, I am taking a page from the ecological activists by asking each of us to act locally to do what we can for each other and ourselves in order to secure the future we want for us and our posterity.

As I look at the world I live in and the world of the past I read about and see in buildings and writings of old, I realize as I did long ago, that the hope for trans generational change comes from the writings of the great thought leaders of human history and the teachings of the individuals all around the world who seek to understand the nature of things through history and science and psychology and to convey that knowledge and understanding to the generations that follow through their acts of teaching.

A billion points of light – my extension of the campaign slogan of President George Bush (the first one). Despite that realities of politics and any disagreement I may have with his policies and processes, the slogan makes a lot of sense. A global movement toward truth, peace, and prosperity is what is needed, and the only way global movements work is for individuals to start them locally.

The strategy I have undertaken is to write and teach and talk and travel and share my ideas and thoughts with others. The tactic of the day is this book, and tomorrow, I will start another one. That is my tactic, but it likely is not yours. You have to choose your own path – that's my rip off of a *Star Wars* rip off of the *Kung Fu* rip off of eastern religion – and they probably got it somewhere else too. Acting locally means that you need to think through the ideas you have, seek to understand the nature of the world you live in, and act to perfect it. That's my rip off of the Jewish tradition of perfecting the Universe – the reason that people exist – and if you will, the meaning of life. I don't buy into the meaning of life part quite yet, but I may get there some day.

You need to find ways to act to save yourself and all of us.

16.4 Who is winning the war?

One thing is for damned sure... it isn't me. But I fight the information war every day, and I'm still standing. I have decided that some things I have to do for myself and I am doing them. Things like writing this book and publishing it give me the direct control over content and format and venues that I cannot have any other way – but it comes at a price. Fewer sales mean that fewer people get to read what I have to say, but at least I get to say it my way.

Who's winning and who's losing? If it isn't obvious by now, today the people of India and China are winning it more and more every day, the people of Western Europe are holding steady, the people of Eastern Europe are making progress with fits and starts, the people of South America appear to be relatively stagnant, the people of the Middle East are losing it and making little progress, the people of Africa are slowly making limited progress but in fits and starts with a very long way to go, the people of Canada are holding steady, and the people of the US are letting it slip away from them.

The individuals of the world are big losers today, while corporations are winning the wars, and criminal organizations are big winners. Governments don't even know they are in these, wars with a few exceptions, while the military Information complex is forming the means to suppress all individuals, the wealthy are prospering as they always do, and the poor are losing it as the great divide between the rich and the poor expands in some places and the poor are winning it slowly where the great divide narrows.

Who can win the information war? We all can. Or we can all lose it. These individual and niche advantages certainly benefit those who have them. As a holder of part of one of those niches, I certainly don't want to give it away, but I would prefer to raise all boats rather than lowering mine to raise yours or raising mine at your expense.

Cooperation is what we all need to win the global information war.

16.5 Conclusion?

Total war used to mean that all of the resources of the State were applied to killing everyone thought to be in any way opposed to the leadership. But in the information age, we have a very different sort of total war. An individualized war in which you and I are engaged in one-on-one warfare with everyone else on the planet at differing levels of intensity.

I have cheated in my long-winded title of this book. It is about the full range of conflict. It is not only about the top end in intensity and frequency that characterizes all out war. And it is not only about the little wars or police actions or insurgencies that last for years or decades after the end of combat operations and in some cases come to take back the victory. It is not only about drug wars in which tens of thousands get killed, or criminal organizations that operate global kidnapping, prostitution, and child pornography rings. It is not only about people disappearing or lies told to promulgate wars, or any of the other enormous range of lower intensity conflicts that effect lots of peoples' lives. It is about all of these and more.

The war I am talking about, in the end, is the war for the hearts and minds of everyone living on Earth. In the end, all war is about extraordinary individual efforts, groups collaborating to take advantage of others, people being fooled and tricked and twisted in to killing others and warped to believe that it is somehow justified. But the war I am talking about is the information war we are all fighting on Earth today, every single one of us. It is the eternal struggle of humanity to rise above the biologically enforced survival instincts that cause us to believe we need to kill to live. The information war we are now engaged in is the truly long war. It will run for longer than I will live and will likely last well beyond the lives of my children and their children. But it is the only war really worth fighting for humanity today. It is the war for peace and prosperity, and you and I together can win it.

Good luck!

Detailed Table of Contents

Printed in the United States
61438LVS00003BA/41

9 781878 109408